Design Guides for Nuclear Power Station

Balance of Plant Mechanical Systems

This book is dedicated to the memory of H. Ray Corbett.

CONTENTS

Preface

After World War II, electric power consumption in the United States increased at a rate of approximately 7% per year. To meet this demand, the Tennessee Valley Authority (TVA) launched an ambitious program of constructing 17 large nuclear power stations, beginning with the Browns Ferry Nuclear Plant in 1966. The portion of the nuclear plants outside of the nuclear steam supply, referred to herein as the "balance of plant" (BOP) was designed by the TVA Division of Engineering Design (DED) in Knoxville, Tennessee.

Rampant inflation approaching 15% annuall during the Carter administration forced the cost of electric power to soar, and the demand for electric power dropped precipitously. At the same time, the disastrous accident at the Three Mile Island Nuclear Generating Station in 1979 turned public opinion against nuclear power. TVA decided not to complete most of the nuclear plants that were under construction. However somewhat belatedly, a consensus may now be forming that nuclear power may be the only hope for society to achieve success against global warming, since it does not produce carbon dioxide.

Realizing that at some point in the future nuclear power might becomes a feasible option again, TVA authorized the highly competent engineers in the DED to produce design guides documenting best engineering practices in designing the BOP of both boiling water and pressurized water reactor nuclear power plants for the benefit of future generations. These design guides reflect the practices employed in the design of the later plants that corrected the many flaws that experience revealed in the earlier designs. These design guides, largely unused for over 40 years, contain a vast amount of useful information. The author was privileged to work alongside the engineers who prepared these design guides for the mechanical discipline and herewith publishes them for the use of posterity.

The Author

Charles F. (Chuck) Bowman, P.E., is the retired President of Chuck Bowman Associates, Inc. (CBA), an engineering consulting firm serving the electric power industry since 1994. Chuck received his BS and MS degrees in Mechanical Engineering from the University of Tennessee, and he is a registered Professional Engineer in Tennessee. CBA specialized in thermal performance analysis of electric power generating cycles and related fields including the design and analysis of heat exchangers, cooling towers, spray ponds, cooling water systems and more. Before forming CBA, Chuck was with the Tennessee Valley Authority (TVA) for 28 years where he supervised the engineering of cooling water systems for the nuclear plants. Prior to his retirement from TVA, he was the Senior Engineering Specialist for thermal performance and cooling water systems in TVA's Corporate Engineering Office. Chuck has served as a consultant to the Electric Power Research Institute, authoring *Alternative to Thermal Performance Testing and/or Tube-Side Inspections of Air-to-Water Heat Exchangers* and the *Turbine Cycle Equipment Engineering Excell Workbook;* has served on ASME committees that authored PTC 23.1 – 1983, Code on Spray Cooling Systems, and on PTC 12.5 - 2000, Single Phase Heat Exchangers; and has served on the American Nuclear Society's ANS-2.21 Working Group, Criteria for Assessing Atmospheric Effects on the Ultimate Heat Sink. Chuck was a contributing editor on the *Marks' Standard Handbook for Mechanical Engineers,* 12th Edition. He is the principal author of *Thermal Engineering of Nuclear Power Stations – Balance of Plant Systems* and *Engineering of Power Plant and Industrial Cooling Water Systems*, both published by the CRC Press, and *Engineering and Design of the Oriented Spray Cooling System.* Chuck has also authored numerous technical papers.

Author's Notes

The design guides contained in this book are exact copies of those prepared, reviewed, and approved by TVA over 40 years ago. Quite obviously, some references to equipment manufacturers and codes and/or industry standards may be out of date. The information is made available to the general public at a nominal cost, and neither the author nor TVA assumes any responsibility for the accuracy of the information contained herein.

The reader may note some variations in terminology between TVA practices and those more common in the industry. For example, TVA denotes the top feedwater heater as Heater No. 1 so that the highest pressure heater is always the same number, regardless of the number of feedwater heaters in the turbine cycle, and the safetly-related cooling water system is denoted as "essential raw cooling water," rather than the more common term "essential service water."

Some of the design guides contained herein make references to other TVA design guides that are not included in this book. The reader may obtain copies of these design guides that may be of interest by contacting the author.

The reader will find additional information in the following books by Charles F. and Seth N. Bowman published by CRC Press:

Thermal Engineering of Nuclear Power Stations – Balance of Plant Systems
Engineering of Power Plant and Industrial Cooling Water Systems.

Chapter 1
Design Guide DG-M2.1.1: Main and Reheat Steam System

1.0 <u>GENERAL</u>

 1.1 SCOPE

 This design guide is an aid for the design of the main and
 reheat steam system of a 3800 megawatt-thermal nuclear plant
 with a light water cooled reactor. Use of this guide permits
 preparation of a draft design criteria for a specific project.
 An engineer with a thorough understanding of the system should
 prepare the design criteria and perform the design calculations.
 Sketches are included for reference and as an aid in preparing
 the design criteria diagram. Appendix A sets forth the necessary
 calculations required for the basic design of the system. Appendix A
 and the sketches should not be a part of the design criteria.

 The discussion of safety-related functions is not within the
 scope of this document.

 1.2 ABBREVIATIONS

 B&W - Babcock and Wilcox Co.

 Btu - British thermal unit

 BWR - Boiling Water Reactor

 CE   - Combustion Engineering Co.

 FW   - Feedwater

 GE   - General Electric Co.

 HP   - High pressure

 HPCV - High pressure control valve

 HPSV - High pressure stop valve

 HR   - Heat rate

 IP   - Intermediate pressure

 kW   - Kilowatt

 kWh - Kilowatt-hour

 LP   - Low pressure

 LPCV - Low pressure control valve

LPSV - Low pressure stop valve

MFP - Main feed pump

MOSV - Motor-operated shutoff valve

MSR - Moisture separator reheater

NSSS - Nuclear steam supply system

OGP - Offgas preheater

OSCV - Overload stop/control valve

PWR - Pressurized water reactor

RCIC - Reactor core isolation cooling

1.3 APPLICABLE CODES AND STANDARDS

All piping between the steam outlet nozzles of the nuclear
steam supply system (NSSS) and the outermost main steam isolation
valves in each steam line leaving the containment shall be
designed and constructed in accordance with the applicable
requirements of ASME Boiler and Pressure Vessel Code, Section III,
Division 1. All remaining piping which is not provided by the
turbine generator contractor shall be designed and constructed
in accordance with the applicable requirements of ANSI B31.1,
"Code for Pressure Piping--Power Piping."

Instrumentation shall be provided as required under ANSI/ASME
PT6-1976, "Test Code For Steam Turbines."

2.0 FUNCTIONAL DESCRIPTION

Figures 1, 2, 3, and 4 are schematics of typical NSSS's for Westinghouse,
B&W, CE, and GE designs. They are provided as an aid to understanding
the information in sections 2.0 through 4.0 of this design guide and
should be consulted as necessary.

In plants utilizing a pressurized water reactor (PWR), the main
steam system delivers steam from the NSSS to meet the requirements
of the turbine generator during all normal plant operating conditions.
In addition, during base load operation the main steam system
supplies steam to the moisture separator reheater (MSR) second
stage reheaters. During low load operation the main steam system
also supplies steam to the main feed pump turbines, turbine seal
system, and auxiliary steam system. If required by the NSSS, the
main steam system supplies pegging steam to feedwater heaters in
order to maintain feedwater temperatures at low loads.

In plants utilizing a boiling water reactor (BWR), the main steam system delivers steam from the NSSS to meet the requirements of the turbine generator during all normal plant operating conditions. In addition, during base load operation the main steam system supplies steam to the MSR second stage reheaters, steam jet air ejectors, and offgas preheaters. During low load or transient operation, the main steam system also supplies steam to the main feed pump (MFP) turbines and auxiliary steam system.

The main steam system boundaries extend from either the steam generator outlet nozzles on a PWR or the outlet of the downstream main steam isolation valve on a BWR, to the inlets of the main turbine stop valves. The steam flows from the NSSS to the high pressure turbine, expands through the turbine, and is taken from the high pressure turbine to the MSR's through the cold reheat piping. In the MSR's, the steam is dried in the moisture-separator section and superheated with two stages of reheat. Steam is carried from the MSR's to the low pressure turbines through the hot reheat piping. After expansion through the low pressure turbines, steam is discharged to the main condenser.

Steam is taken from the NSSS to the main steam inlet header. Flow nozzles, if required (not required by Babcock and Wilcox (B&W) NSSS), are supplied by the NSSS contractor. These flow nozzles plus the main steam isolation valves are located in the lines to the header. The main steam inlet header supplies steam to the high pressure turbine, MSR second stage reheaters, MFP turbine high pressure steam inlets, and auxiliary steam system. In addition, on a BWR the main steam inlet header supplies steam to the offgas preheaters, steam jet air ejectors, and radwaste system. (Steam may also be supplied from the header to the first high pressure turbine extraction point through an overload stop/control valve, at the turbine generator manufacturer's option. This permits operating the unit beyond the turbine guarantee load. There shall be no valve between the overload stop/control valve and the turbine. The normal valve arrangement, that is, isolation valve and non-return valve, shall be provided between the overload valve and the no. 1 heater.)

A turbine bypass system capable of taking steam from the main steam inlet header directly to the condenser is provided for unit startup and transients, including trips.

Piping may be provided to transport steam between the main steam headers of the units via the auxiliary steam system during startup of either unit when the other unit is operating.

3.0 SYSTEM DESIGN REQUIREMENTS

3.1 STARTUP

Main steam piping shall be adequately warmed prior to admission of steam to the turbine generator. Upon opening of the main steam isolation valves, the main steam system shall be capable of supplying steam to the high pressure turbine and MSR second stage reheater. Steam will be taken from the main steam inlet header, when pressure is adequate, for the main turbine and the auxiliary steam system. The system shall be capable of bypassing excess steam from the NSSS directly to the condenser.

Prior to opening of the main steam isolation valves on a PWR unit, turbine seal steam and condenser deaerating steam is provided from the auxiliary steam system.

Prior to opening of the main steam isolation valves on a BWR unit, steam for the steam jet air ejectors, the offgas preheaters, and condenser deaerating is provided from the auxiliary steam system.

3.2 SHUTDOWN

During normal unit shutdown, the main steam system shall supply steam to the turbine generator, turbine bypass, and MFP turbines to withdraw the unit from service.

3.3 HOT STANDBY

With the unit on hot standby, the main steam system shall bypass steam to the main condenser. Steam shall also be supplied to the MFP turbines. If required by the NSSS, pegging steam shall be supplied to the feedwater heaters.

3.4 POWER OPERATION FROM PARTIAL LOAD TO MAXIMUM ANTICIPATED POWER LEVEL

During power operation, the main steam system shall supply steam to the high pressure turbine and the MSR second stage reheater. Whenever possible, the MFP turbines will utilize steam from the MSR outlet. However, at low loads or upon loss of a MFP, the MFP turbines will supplement this extraction steam with high pressure steam from the main steam inlet header as necessary. If required by the NSSS during low load operation, pegging steam shall be supplied to the feedwater heaters.

Either unit of a two-unit plant shall also provide steam to the auxiliary steam system to permit startup of the other unit.

4

3.5 TRANSIENT CONDITIONS

The system shall satisfy the steam requirements of the unit under transient conditions, including the following:

a. Design plant load changes, i.e.,

   1. Step load changes of ±10 percent throughout the load range of 15 to 100 percent.

   2. Ramp load changes throughout the load range of 15 to 100 percent, corresponding to the capability of the NSSS.

b. Simultaneous turbine and reactor trip with normal operation of the turbine bypass system to dissipate heat.

c. Complete loss of load and failure of the turbine bypass system to operate. The main steam safety valves are assumed to open and credit is taken for a reactor trip.

d. Full load rejection or turbine trip with normal operation of the turbine bypass system and reactor runback to prevent a reactor trip.

4.0 DESIGN INFORMATION

4.1 PERFORMANCE

4.1.1 Pressures and Flows

The design pressure of the piping to the MFP turbine low pressure steam admission shall be the same as the MSR vessels. The design pressure of the remaining main steam piping shall be the same as the steam generators on a PWR (the reactor pressure vessel on a BWR). These pressures are obtained from the turbine generator contractor and the NSSS contractor, respectively.

Required flow rates through the steam supply piping are obtained from the NSSS contractor, turbine generator contractor, and the contractors of any other equipment which uses main or reheat steam.

4.1.2 Electrical Power

Components of the main steam system shall be powered from normal station service power supplies. No component shall be required to function on loss of offsite power. Therefore, no interfacing with the diesel generators is needed.

4.2 COMPONENTS

4.2.1 Main Steam Piping

Steam for the high pressure turbine shall be taken from the main steam inlet header through four parallel steam lines. Main steam stop valves followed by turbine control valves and all necessary interconnecting piping between the main steam stop valves and the high pressure turbine shall be provided by the turbine generator contractor. Provisions shall be made in the high pressure turbine and the inlet piping for pressure transducers for indication of turbine throttle and first stage pressures.

4.2.2 Cold Reheat Piping

The turbine generator contractor shall provide all interconnecting piping between the high pressure turbine exhaust and the MSR main steam inlets. Extractions may be provided in the piping for supply of cold reheat steam to the heating, ventilating, and air conditioning system, steam seal evaporator, and radwaste evaporator.

4.2.3 Hot Reheat Piping

The turbine generator contractor shall provide all interconnecting piping between the MSR outlets and the low pressure turbine inlets. Each steam line shall be provided with a reheat stop valve and an intercept valve. Provisions shall be made for pressure transducers for indication of low pressure turbine inlet pressure. Extractions shall be provided in the piping or the MSR to supply hot reheat steam to the MFP turbines.

4.2.4 Main Feed Pump Turbine Piping

The MFP turbines shall have a low pressure steam admission utilizing hot reheat steam for normal operation and a high pressure steam admission utilizing main steam for low load operation and contingency conditions. The MFP turbine contractor shall provide an automatic stop valve for each steam inlet and governor control valves as required for efficient operation of the turbine. The high pressure stop valve shall be provided with an arrangement to permit steam line warming and condensate removal when the high pressure steam admission is not in use.

The steam piping to each low pressure admission shall be
provided with a flow element, a motor-operated isolation
valve, and a check valve. The steam piping to each high
pressure admission shall be provided with motor-operated
isolation valve. One additional flow element shall be
located in the steam line from the main steam inlet header
serving the high pressure steam admission of all three MFP
turbines.

4.2.5 Turbine Bypass System

The turbine bypass system shall have a total flow capability
as determined by the NSSS contractor. Turbine bypass
control valves shall modulate steam flow to the condenser
when the bypass system is in operation. (The bypass control
valves shall fail safe closed to prevent uncontrolled
bypass of steam to the main condenser.) A shutoff valve
is provided upstream of each turbine bypass control valve
to allow for unit operation upon partial control valve
failure or impairment. Temperature detectors are located
downstream of each bypass control valve for main control
room indication of valve leakage.

The condenser internally distributes the steam from the
turbine bypass valves as required between the high pressure
(HP), intermediate pressure (IP), and low pressure (LP)
zones. (On PWR's with a turbine bypass capacity of greater
than about 40 percent, some of the bypass valves may discharge
to atmosphere.) The condenser shall be sized to pass the
bypass system steam flow capability at an inlet pressure of
250 psig. The condenser shall be capable of accepting the
dump steam in addition to the remaining steam coming through
the turbine, without exceeding the maximum backpressure
allowed by the turbine generator contractor at the maximum
cooling water temperature and the maximum permissible
differential pressure between any two adjacent pressure
zones (as determined by either the turbine generator or
condenser contractor). The condenser shall be designed so
that the temperature at the turbine-condenser interface
does not exceed approximately 175°F during steam dump.
(Exact value to be provided by the turbine generator contractor.)

4.2.6 Auxiliary Steam System

The auxiliary steam system will be common to both units.
When one unit is out of service the main steam inlet header
of the other unit will supply steam to the auxiliary steam
system. Auxiliary boilers are provided for periods when
both units of the plant are not in operation as described
in the auxiliary steam system design criteria.

4.2.7 Emergency Feedwater System (PWR Only)

Steam for the emergency feedwater turbine shall be taken
from the main steam piping outside of the containment
upstream of the main steam isolation valves as described in
the emergency feedwater system design criteria.

4.2.8 Reactor Core Isolation Cooling System (BWR Only)

The reactor core isolation cooling system shall be provided
by the NSSS contractor. Steam for the reactor core isolation
cooling pump turbine shall be taken from the main steam
piping within the containment upstream of the main steam
isolation valve.

4.2.9 Valves (PWR Only)

4.2.9.1 Main Steam Isolation

The main steam isolation valves will be provided as
described in the steam generator isolation and relief
system design criteria and the primary containment isolation
system design criteria. The main steam isolation valves
shall be capable of shutoff against flow in either direction.
A bypass shall be provided around the main steam isolation
valves for use during steam line warming and unit startups.

4.2.9.2 Main Steam Relief

Main steam relief valves provided by the NSSS contractor
will be as described in the steam generator isolation and
relief system design criteria. The main steam relief
valves shall be located outside of the containment upstream
of the main steam isolation valves. The main steam
relief valves shall discharge to the atmosphere.

4.2.9.3 Atmospheric Dump Valves

Power-operated atmospheric dump valves shall be provided
as described in the steam generator isolation and relief
system design criteria to allow for cooldown of the steam
generators when the main steam isolation valves are
closed or the condenser is not available as a heat sink.
The atmospheric dump valves shall be located outside of
the containment upstream of the main steam isolation
valves.

4.2.9.4   MFP Turbine Isolation

The MFP turbine motor-operated isolation valves shall be capable of opening against a maximum differential pressure equal to the design pressure. Each of these isolation valves can be controlled by the operator from the main control room.

4.2.9.5   MSR Valves

The turbine generator contractor shall provide a control valve at the MSR second stage reheater steam inlet. A stop valve shall be provided at the second stage reheater inlet with a bypass valve for warming prior to startup.

The turbine generator contractor shall provide relief valves for the MSR as required by design. These relief valves shall be designed for full-vacuum operation. Temperature detectors shall be located downstream of each relief valve for main control room indication of valve leakage. Each relief valve shall have a separate discharge line to atmosphere through the turbine building roof with no manifolding or interconnections.

4.2.10   Valves (BWR Only)

4.2.10.1 Main Steam Isolation

The main steam isolation valves shall be capable of shutoff against flow in either direction. A bypass shall be provided around the main steam isolation valves for use during steam line warming and unit startups.

4.2.10.2 Safety Relief Valves

Dual function safety relief valves shall be provided by the NSSS contractor. The safety function provides protection against overpressure of the reactor primary system, and the relief function provides power-actuated valve opening to depressurize the reactor primary system. The safety relief valves are located within the containment, and discharge directly to the pressure suppression pool.

4.2.10.3 MFP Isolation Valves

The MFP turbine motor-operated isolation valves shall be capable of opening against a maximum differential pressure equal to the design pressure. Each of these isolation valves can be controlled by the operator from the main control room.

4.2.10.4 MSR Valves

The turbine generator contractor shall provide a control valve at the MSR second stage reheater steam inlet. A stop valve shall be provided at the second-stage reheater inlet with a bypass valve for warming prior to startup.

The turbine generator contractor shall provide relief valves for the MSR as required by design. These relief valves shall be designed for full-vacuum operation. Temperature detectors shall be located downstream of each relief valve for indication of valve leakage. Each relief valve shall have a separate discharge line to the condenser with no manifolding or interconnections.

4.2.11 Piping (PWR and BWR)

Piping design shall include considerations to insure against excessive reactions on all equipment connections and the containment.

Low point drains shall be either orificed or trapped and should operate continuously. Drain pots (or enlarged pipe sections) shall be provided near the drain connection with a level switch for main control room annunciation of drain malfunction.

Low point drains shall also be equipped with an automatic bypass valve for use during startup, low load operation, and trap failure.

The startup drains shall be sized to pass the required condensate flow for warmup of the main turbine with static head as the only driving potential.

In the BWR and B&W NSSS, there shall be provisions for warming reheater tube bundles from the auxiliary steam system to prevent corrosion of the carbon steel tubes during shutdown.

4.2.12 Instrumentation

In addition to the instrumentation required to perform the logical functions described herein, additional instrumentation shall be provided to measure system temperatures and pressures required to satisfy the input needs of the balance of plant computer system, provide test information as required under ANSI/ASME PTC6-1976, and give additional information regarding operation of the system equipment.

If the output of the sensing elements is not required for plant computer or logic functions, output display on local panels shall be sufficient.

All pressure monitoring connections shall be equipped to facilitate on-line instrument calibration without disturbing station instrumentation.

4.2.13 Insulation

All piping and equipment whose temperature may exceed 150°F shall be insulated to provide surface temperatures that do not exceed 125°F with ambient air at 80°F and 50 ft/min velocity.

4.3 OPERATING LOGIC

4.3.1 Startup

The unit startup will be done by manually starting the condensate hotwell pumps and condensate booster pumps. Steam line warming will be done by opening all bypass valves in piping low point drains. The turbine steam seal system will be placed into operation and the main turbine and MFP turbines will be warmed using the turbine drains and vents. Excess steam will be taken to the condenser by the turbine bypass system. Details of the turbine startup will be provided by the turbine contractors.

When steam pressure exceeds approximately 600 psia, the first MFP turbine will be placed into service using steam from the main steam inlet header. The turbine generator will be rolled, brought up to speed, and synchronized to begin power operation. The second and third MFP turbines will be started and placed into service at approximately 25 percent and 50 percent rated steam flow, respectively. Turbine startup drains and vents will be closed, and low point drains will be placed into automatic operation (bypasses normally closed) above approximately 15 percent load.

4.3.2 Low Point Drains

All bypass drains shall be controlled from a single switch in the main control room. Capability for controlling individual bypass drain valves shall also be provided locally.

APPENDIX A

DESIGN CALCULATIONS

FOR

MAIN AND REHEAT STEAM SYSTEM (NUCLEAR)

1.0 PURPOSE

This appendix outlines the calculations necessary for the basic design of the main steam system. In addition to being an aid to the engineer responsible for the system design, this appendix sets forth a format for calculations. This ensures that system design will be consistent among TVA plants and that the calculations are a convenient source of reference.

2.0 BASIC INFORMATION

The following information is necessary to perform the main steam system calculations:

2.1 From the NSSS Contractor

    a. The steam outlet pressure and flow should be obtained at 100 percent of NSSS rating. The moisture content of the steam (or the degree of superheat for a B&W NSSS) should be provided at the steam outlet for these design points.

    b. The turbine bypass system capability and maximum allowable pressure drop from the NSSS steam outlet to the turbine bypass valves.

    c. Design pressure of the steam generator (PWR only) or the reactor pressure vessel (BWR only).

2.2 From the Task Force Plant Layout and Piping Group

Piping isometric sketches showing lengths between changes in pipe diameter or branches in piping and approximate location of valves and fittings. The isometric should show the locations of the steam outlet nozzles on the NSSS, the main steam inlet header, and the location of the turbine generator main steam stop valves. The plant layout and equipment group should refer to the Design Criteria Diagram for Main Steam Piping Requirements. The main steam system sketch is included in this design guide as an aid in developing the design criteria diagram.

2.3 From the Main Feed Pump Turbine Contractor

Required MFP turbine steam flows at 105 percent, 100 percent, 75 percent, 50 percent, and 25 percent load with three MFP turbines in operation. In addition, on a BWR the MFP turbine steam flow required to deliver 115.5 percent flow to the NSSS with three MFPs in operation shall be provided.

2.4 From the Nuclear Steam Generation and Equipment Group

A recent contract for main steam piping should be used to obtain the cost on a dollar-per-pound basis. These contracts should be available through the General Procurement Sections. Piping costs should be escalated according to expected cost trends.

2.5 From the Office of Power

Values for changes in net plant heat rate (in $/Btu/kWh) and generating capability (in $/kW) should be provided in order to optimize the main steam system piping.

3.0 MAIN STEAM LINE SIZING

3.1 The turbine generator shall have a maximum guaranteed rating corresponding to the NSSS 100 percent rating (hereafter referred to as 100 percent load). For the purpose of specifying the steam inlet conditions of the turbine generator on a PWR, a pressure drop of 25 psi should be assumed from the steam generator outlet nozzles to the main turbine stop valve. Since principal piping on a BWR up to and including the main steam isolation valves is provided by the NSSS contractor, a pressure drop of 15 psi should be assumed from these valves to the main turbine stop valves. Moisture (or degree of superheat for a B&W NSSS) should be calculated at the turbine inlet assuming no change in enthalpy between the NSSS steam outlet and the main turbine stop valve.

3.2 The optional sizes for the main steam piping will consider the cost of material and the effects of the resulting pressure drop on heat rate and generating capability. It is suggested that a previous plant design be used for a first estimate of steam line sizes. Piping wall thickness should be estimated, pressure drop through the main steam system calculated, and the net plant heat rate and generating capability determined as stated in sections 3.2.1 through 3.2.6 in the appendix.

13

3.2.1 Piping in sizes larger than 24 inches diameter (ASTM A-155 KCF70) is rolled from plate and longitudinally welded to nominal outside dimensions. The following formula should be used to calculate the minimum required wall thickness in inches:

$$t = \frac{1.005\ PD}{2(S + Py)} + A$$

where:

P = Piping internal design pressure, equal to the design pressure of the steam generator on PWR plants or the reactor pressure vessel on BWR plants (psig).

D = Nominal outside diameter of the pipe (inches)

S = Maximum allowable stress (17,500 psi for ASTM A-155 KCF70)

y = Coefficient, as stated in ANSI B31.1, Table 104.1.2.A (0.4 for temperatures 900°F and below)

A = Additional thickness to provide for corrosion and/or erosion (0.08 in.).

This equation differs from the equation given in ANSI B31.1, Para. 104.1.2 by a factor of 1.005 to allow for a piping O.D. tolerance of one-half percent.

The specified pipe wall thickness will be greater than the minimum required wall thickness in order to allow for machining and O.D. tolerances of the fittings and weld fit-up. The project will make the actual calculation of piping wall thickness; however, the specified wall thickness can be estimated by increasing the minimum required wall thickness by 115 mils so that the internal pipe diameter "d" used in DG-M2.11, "Pressure Drop Calculations for Steam and Condensate Piping Fittings," is determined by:

$$d = D\ 2\ (t + 0.115)$$

3.2.2 DG-M2.11 should be used to obtain pressure drops in the main steam piping at 100 percent load. Note that Table 2 of DG-M2.11 is not applicable for ASTM A-155 piping.

Pipe sizes should be chosen to limit steam velocities to a maximum of 150 ft/sec. The piping isometric sketches from the Task Force Plant Layout and Piping Group should be used to obtain piping lengths, numbers of fittings, etc. The calculations should consider pipe friction loss and head losses in fittings, flow nozzles, and valves. A pressure drop of 3 psi should be assumed for the main steam isolation valves. On PWR's, a pressure drop of 5 psi should be assumed for flow restrictors.

3.2.3 Once the pressure drop through the main steam piping has been calculated, the turbine initial pressure should be determined by subtracting the pressure drop from the NSSS outlet pressure. Moisture (or degree of superheat for a B&W NSSS) should be calculated assuming no change in enthalpy between the NSSS steam outlet and the main turbine stop valve. Changes in heat rate should be estimated using the following formula:

$$HR(2) = HR(1) \ [1 + (0.15/1000) \ (1000 - P)] \ (1 + M/100)$$

where:

HR(1) = Turbine generator heat rate based on steam inlet conditions of 1000 psia with dry saturated steam (Btu/kWh). This is found with the above equation by substituting in the appropriate value for P and M from the guaranteed heat balance using the guaranteed heat rate for HR(2).

P = Turbine inlet pressure (psia).

M = 0.265 increase in heat rate for each one percent increase in moisture content of the steam entering the main turbine stop valves, i.e., for a moisture content of 0.4 percent M = (0.265)(0.4). M = -0.3 percent change in heat rate per each 10°F of superheat for a B&W NSSS, i.e., for 45°F of superheat M = -(0.3)(4.5).

HR(2) = Turbine generator heat rate corrected for initial pressure and moisture content (or superheat) of the steam.

3.2.4 The change in generating capability of the turbine generator should be estimated using the corrected values of heat rate and the following relationship:

$$KW(2) = KW(1) \ \frac{HR(1)}{HR(2)}$$

where:

KW(1) = Guaranteed turbine generator capability in kW.

HR(1) = Guaranteed turbine generator heat rate.

HR(2) = Corrected turbine generator heat rate.

KW(2) = Corrected turbine generator capability.

3.2.5   The investment cost of the piping should be obtained using the length of piping, the piping wall thickness, and a density of steel of 0.284 lb/cu in. The higher investment cost of the larger diameter piping should be evaluated against the savings in operating cost due to improved heat rate and greater generating capability of the unit. The savings in operating costs are given by the corrected values of heat rate (Btu/kWh) and turbine generator capability (kW) and using the evaluation factors for each of these changes.

3.2.6   The calculations described in sections 3.2.1 through 3.2.5 should be done for a number of alternate piping sizes until an optimum diameter is obtained. Piping diameters are normally chosen in even outside diameters (i.e., 34 inches, 36 inches, 38 inches). Piping flexibility may finally determine the final selection of pipe diameters. For this reason, it would be helpful to arrange the alternates in order according to their evaluated costs. Then, if the most economical alternate is unattractive because of flexibility or other reasons, the next most economical alternate and its relative cost can be easily compared.

3.3   Once the main steam line sizes have been selected, the pressure in the main steam inlet header should be calculated for use in other calculations.

4.0   TURBINE BYPASS LINE SIZING

The pressure drop from the NSSS steam outlet to the main steam inlet header should be subtracted from the maximum allowable pressure drop from the NSSS steam outlet to the turbine bypass valves. The remaining pressure drop should be used to size the turbine bypass piping, limiting steam velocities to less than 350 ft/s. Piping in sizes larger than 24 inches should be ASTM A-155 KCF70 with inside diameters calculated as stated in section 3.2.1. DG-M2.11 "Pressure Drop Calculations for Steam and Condensate Piping Fittings" should be used.

5.0   MFP TURBINE LINE SIZING

The steam line from the main steam inlet header to the MFP turbine high pressure admission should be sized for an approximate pressure drop of 50 psi from the NSSS steam outlet to the MFP turbine high pressure admission (the pressure drop of 50 psi includes the pressure drop from the NSSS steam outlet to the main steam inlet header). Pipe sizes should be chosen to limit steam velocities to 150 ft/s.

The steam line from the hot reheat piping or MSR to the MFP turbine low pressure admission should be sized for a pressure drop of 3 percent of the crossover pressure. (A pressure drop of up to 5 percent can be used if the steam line is found to be too large with a smaller pressure drop.)

Steam velocities should be limited to a maximum of 150 ft/s. Piping in sizes 24 inches and smaller should be ASTM A-106, Grade B. DG-M2.11 "Pressure Drop Calculations for Steam and Condensate Piping Fittings" should be used.

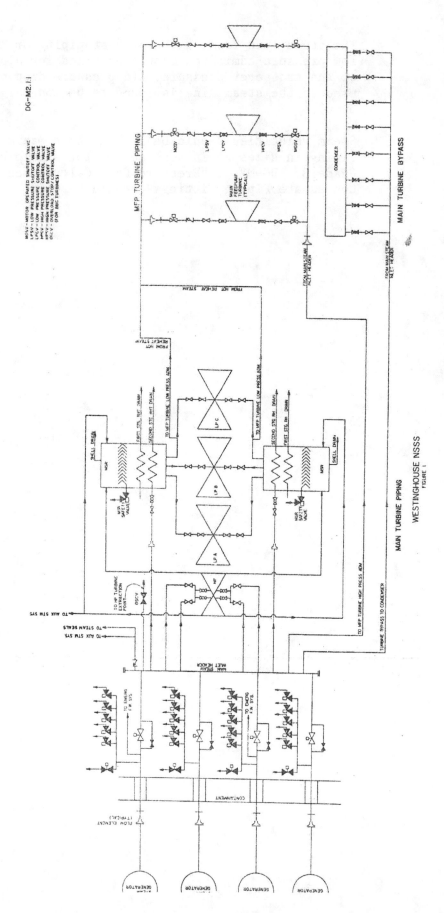

WESTINGHOUSE NSSS
FIGURE 1

18

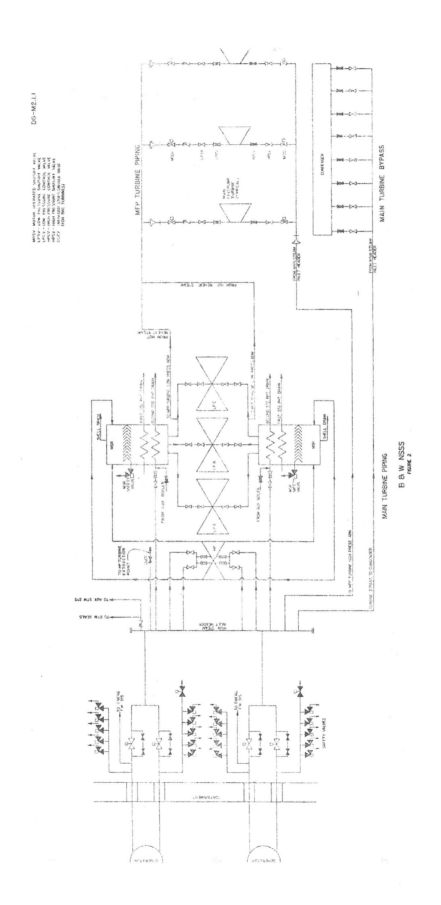

B & W NSSS
FIGURE 2

19

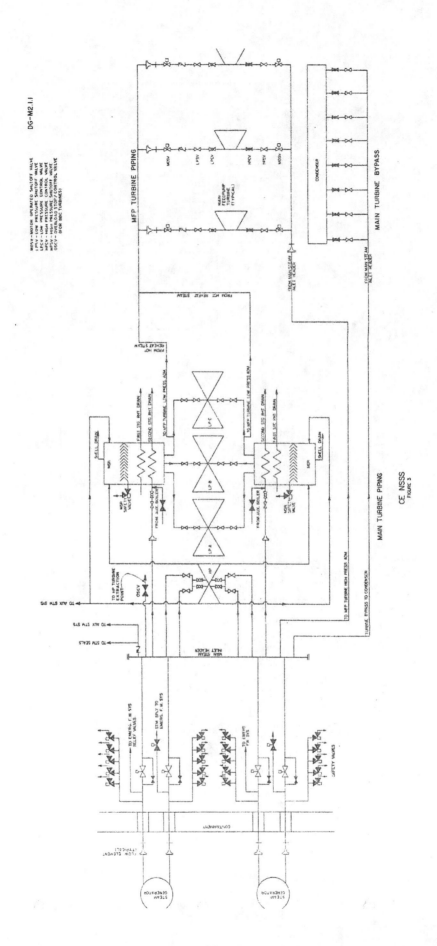

DG-M2.1.1

MDSV - MOTOR OPERATED SHUTOFF VALVE
LPSV - LOW PRESSURE SHUTOFF VALVE
LPCV - LOW PRESSURE CONTROL VALVE
HPCV - HIGH PRESSURE CONTROL VALVE
HPSV - HIGH PRESSURE SHUTOFF VALVE
OSCV - OVERLOAD STOP / CONTROL VALVE
(FOR BBC TURBINES)

MFP TURBINE PIPING

MAIN TURBINE PIPING

CE NSSS
FIGURE 3

20

Chapter 2
Design Guide DG-M2.2.1: Main Feedwater System

1.0 <u>GENERAL</u>

    1.1 SCOPE

    This design guide is an aid for the design of the non safety-related main feedwater system of a 3800 megawatt-thermal nuclear plant with a light water cooled reactor. Statements related to system boundary extending beyond the first seismic anchor (normal TVA-NSSS interface) are intended to define system functional (performance) requirements. Physical design requirements for piping, valves, hangers, etc., between the first seismic anchor and the reactor or steam generator are contained in other design guides.

    Use of this design guide permits preparation of a draft design criteria for a specific project. An engineer with a thorough understanding of the system should prepare the design criteria and perform the design calculations. Appendix A includes a schematic and the necessary calculations for the basic system design.

    1.2 ABBREVIATIONS

        B&W   - Babcock and Wilcox, Inc.

        BHP   - Brake horsepower

        BWR   - Boiling Water Reactor

        CBP   - Condensate Booster Pump

        CE   - Combustion Engineering, Inc.

        FPBV   - Feed Pump Bypass Valve

        GE   - General Electric

        H   - Total head (feet)

        HWP   - Hotwell Pump

        IVBV   - Isolation Valve Bypass Valve

        MFP   - Main Feedwater Pump

        N   - Pump speed (rpm)

        NBR   - Nuclear Boiler Rating

        NPSH   - Net Positive Suction Head

NSSS   - Nuclear Steam Supply System

PCCV   - Positive Closing Check Valve

PWR   - Pressurized Water Reactor

Q   - Pump capacity (gpm)

S   - Suction specific speed

SG   - Steam Generator (PWR)

SLCBV   - Startup Level Control Bypass Valve

SLCV   - Startup Level Control Valve

} BWR only

T-G   - Turbogenerator

1.3 APPLICABLE CODES AND STANDARDS

All components which contain the system pressure shall be designed and constructed in accordance with the applicable requirements of the ANSI B31.1, "American National Standard Code for Pressure Piping--Power Piping," and/or the ASME Boiler and Pressure Vessel Code, Section VIII, Division 1. Feedwater heaters shall be designed and constructed in accordance with the Heat Exchange Institute's Standard for Closed Feedwater Heaters. Heater tubes shall conform to ASTM A688, "Standard Specification for Welded Austenitic Stainless Steel Feedwater Heater Tubes," with allowable stress values as defined in ASTM A249, "Standard Specification for Welded Austenitic Steel Boiler, Superheater, Heat-Exchanger, and Condenser Tubes."

The system pumps shall comply with applicable sections of the standards of the Hydraulic Institute.

2.0 FUNCTIONAL DESCRIPTION

The feedwater system shall deliver water from the suction of the main feed pumps (MFP's) to the Nuclear Steam Supply System (NSSS) feedwater inlets at the flow, water chemistry, temperature, and pressure required by the NSSS during all normal plant operating conditions.

Refer to figure A-5 for the following discussion.

The feedwater system boundaries extend from the MFP suction to the NSSS feedwater inlets. Condensate is delivered to the MFP suction from the condensate system and from the heater drain system (No. 3 heater drain pump discharge) in sufficient quantity and with sufficient net positive suction head to ensure that the MFP's will be capable of meeting the requirements of the NSSS.

The MFP's discharge into a common header from which the feedwater passes through two stages of feedwater heating in three parallel streams. (The heaters are numbered sequentially in order of decreasing shell side pressure, and heater streams are designated by letters; the highest pressure heaters are designated 1A, 1B, and 1C.) The feedwater leaving the No. 1 heaters is combined into a single, full-flow line to ensure proper mixing and minimize enthalpy variations between the NSSS feedwater inlets.

For PWR units, feedwater flows to each steam generator through individual feedwater lines, each containing a control valve and a flow nozzle. Containment and steam generator isolation are provided by two automatic isolation valves located in the seismic portion of each feedwater line just outside the containment. One or two check valves are provided downstream of the isolation valves.

For BWR units, feedwater flows to the reactor pressure vessel through two feedwater lines, each containing a flow nozzle for feedwater flow measurement. Isolation of the containment and reactor vessels is provided by an automatic isolation valve and a positive-closing check valve outside the containment, and a check valve inside the containment. A shutoff valve is provided for maintenance purposes in each feedwater line between the containment isolation check valve and the reactor pressure vessel.

3.0   SYSTEM DESIGN REQUIREMENTS

3.1   STARTUP

Under startup conditions, the feedwater system will be designed so that water from the condensate system can be bypassed around the main feed pumps, through the high pressure feedwater heaters, and back to the main condenser. This recirculating water flow is required for system cleanup to ensure that the quality of the feedwater is within the limits required by the NSSS. Following the cleanup operation, the system will be capable of delivering water directly to the NSSS while bypassing the main feed pumps. As NSSS heatup progresses, the pressure at the NSSS feedwater inlet will increase until it exceeds the head capacity of the condensate system pumps. Prior to reaching this point, one turbine-driven main feed pump will be placed in service and the MFP bypass line will be isolated.

23

3.2 SHUTDOWN

During normal unit shutdown, the feedwater system will supply water to the NSSS for removal of residual heat until an auxiliary cooling system becomes available.

3.3 HOT STANDBY

With the unit on hot standby, the feedwater system (with main feed pump(s) operating) will supply water to the NSSS as required.

3.4 POWER OPERATION FROM PARTIAL LOAD TO MAXIMUM ANTICIPATED POWER LEVEL

During power operation, the feedwater system shall deliver feedwater to the NSSS as required by the NSSS-supplied feedwater control system.

3.5 TRANSIENT CONDITIONS

The system shall satisfy the feedwater requirements of the NSSS under transient conditions including the following:

a. (PWR only) Design plant load changes, i.e.,

    1. Step load changes ±10 percent throughout the load range of 15 to 100 percent.

    2. Ramp load changes throughout the load range of 15 to 100 percent corresponding to the capability of the NSSS.

b. (BWR only) Design plant load changes, i.e.,

Load Range (% NBR)	Max. Load Step Without Reactor Scram	Max. Load Ramp Without Reactor Scram
0-25	10%	±30 MW/min
25-50	$\pm10\%$	$\pm0.25\%$/sec (automatic flow control)
50-75	$\pm18\%$	$\pm0.5\%$/sec for up to 10% total load change
		$\pm0.25\%$/sec for larger changes
75-100	$\pm25\%$	$\pm0.5\%$/sec for up to 10% total load change
		$\pm0.25\%$/sec for larger changes

c. Simultaneous turbine and reactor trip with normal operation of the turbine bypass system to dissipate heat.

d. Complete loss of load and failure of the turbine bypass system to operate. (The main steam safety valves shall be assumed to open and credit for a reactor trip shall be taken.)

e. Loss of one stream of feedwater heaters without reducing load to less than 100 percent of NSSS rating.

f. Loss of one main feed pump without reducing load to less than 100 percent of NSSS rating.

3.6 MISCELLANEOUS CONDITIONS

The condensate system shall provide water for initial fill and wet layup through the MFP bypass line and auxiliary control valve (PWR only) or startup level control bypass valve (SLCBV) (BWR only).

Flooding of the fuel pool is to be accomplished through the startup level control valve also.

4.0 SPECIFIC CRITERIA

4.1 PERFORMANCE CRITERIA

4.1.1 Main Feed Pumps

4.1.1.1 Each main feed pump shall be designed for best efficiency at 33-1/3 percent of the feedwater system design flow [based on turbine generator guaranteed heat balance (corresponds to NSSS guaranteed heat balance)] while producing total head required by the NSSS with three hotwell pumps, three condensate booster pumps, all feedwater heaters, and all heater drain pumps in service with maximum pressure drop across the demineralizers.

4.1.1.2 Two main feed pumps with three hotwell pumps, three condensate booster pumps, all feedwater heaters, and all heater drain pumps in service, with maximum pressure drop across the demineralizers, shall provide 100 percent of the feedwater system design flow with the required pressure at the feedwater inlet nozzles of the NSSS.

4.1.1.3 Three main feed pumps with two hotwell pumps, two condensate booster pumps, all feedwater heaters, and all heater drain pumps in service, with maximum pressure drop across the demineralizers shall provide 100 percent of the feedwater system design flow with the required pressure at the feedwater inlet nozzles of the NSSS.

4.1.1.4   For PWR units, three main feed pumps with three hotwell
           pumps, three condensate booster pumps, all feedwater
           heaters, all heater drain pumps in service, and maximum
           pressure drop across the demineralizer shall provide 105
           percent "feedwater system design flow" with the required
           pressure at the inlet nozzle on the steam generator.

           For BWR units, three main feed pumps with three hotwell
           pumps, three condensate booster pumps, all feedwater
           heaters, all heater drain pumps in service, and maximum
           pressure drop across the demineralizer shall provide 115.5
           percent (1.1 x 105 percent continuous flow for transient
           stability) of the feedwater system design flow with the
           required pressure at the reactor feedwater sparger inlet
           nozzle.

4.1.1.5   Three main feed pumps with three hotwell pumps, three
           condensate booster pumps, and all heater drain pumps in
           service, and maximum pressure drop across the demineralizers
           shall provide 100 percent of the feedwater system design flow
           with the required pressure at the feedwater inlet nozzles of
           the NSSS with one stream of feedwater heaters out of service.

4.1.1.6   The main feed pumps shall be designed to provide feedwater
           as required for transient operation of the NSSS system (see
           section 3.5).

4.1.2   High Pressure Feedwater Heaters

4.1.2.1   Three streams of high pressure feedwater heaters with three
           streams of low pressure heaters in service shall be capable
           of providing 100 percent of the feedwater system design flow
           to the NSSS at the final feedwater temperature required by
           the NSSS contractor.

4.1.2.2   The heaters shall have a 5°F terminal temperature difference
           and 10°F approach drain coolers.

4.1.2.3   The maximum channel side differential pressure including
           inlet and discharge nozzles shall not exceed 25 psi total
           for both high pressure heater stages at feedwater system
           design flow.

4.1.2.4   Two streams of high pressure feedwater heaters shall be
           capable of passing 100 percent of the feedwater system
           design flow when one stream of feedwater heaters is isolated.

26

4.1.3 Power Supply

Components of the main feedwater system shall be powered from normal station service power supplies. No component shall be required to function on loss of offsite power and no interfacing with the diesel generators is needed.

4.2 COMPONENT CRITERIA

4.2.1 Main Feed Pumps

4.2.1.1   The three main feed pumps shall be horzontal, turbine-driven, single-stage, centrifugal pumps. The pump casing shall be a double-volute or diffuser type, made of alloy cast steel with a nominal 13 percent chromium and 4 percent nickel content. The impeller shall be of alloy cast steel either with a nominal 13 percent chromium and 4 percent nickel or 11-13 percent minimum chromium.

4.2.1.2   Each pump shall have stuffing boxes of the packingless type (throttle bushing, floating rings, mechanical seals, etc.). The pumps shall be constructed such that the seals will be subjected to suction pressure only. Each pump shall be furnished with complete injection controllers (if required) and accessories, including filters and stainless steel piping between filters and pump. Each pump shall be provided with the necessary drain, vent, and cooling connections.

4.2.1.3   Each pump will be supplied with adequate lubricating oil from the turbine driver during starting up, shutting down, turning gear, and normal operation. The lubricating system supplied by the main feed pump turbine contractor shall include the necessary oil pump, filter, relief valve, pressure switches, oil reservoir with level indicator, oil coolers, thermometer in oil cooler oil discharge line, and piping to connect to the pump lubrication system. (See DG-M2.19.1, "TURBINE GENERATOR HEAT CYCLE--Main Feed Pump Turbine System" for details).

4.2.1.4   Each main feed pump shall be supplied with a minimum flow recirculation system consisting of a flow element in the pump suction piping, automatic controls, and a pressure breakdown valve. The pressure breakdown valve shall be the modulating type and shall fail open on loss of control air or control signal. The main feed pump recirculation system shall be preliminarily sized for approximately 30 percent of the "feedwater system design flow" per pump. Final sizing parameters will be supplied by the feed pump contractor.

4.2.1.5   Each pump shall be furnished with a force lubricated,
           segmented, pivot shoe thrust bearing (Kingsbury type)
           capable of carrying equal thrust loads in either direction
           to prevent axial metallic contact under any operating
           condition.

4.2.1.6   The design pressure of the main feed pumps shall be equal
           to the maximum discharge pressure of the main feed pumps
           at minimum recirculation flow, rated speed, and with the
           maximum suction head from the hotwell and condensate booster
           pumps.

4.2.2   High Pressure Feedwater Heaters

4.2.2.1   For PWR units, each heater shall be a closed, horizontal,
           U-tube type with provisions for shell pulling to provide
           access to tube bundles for maintenance. Tubes shall be
           Type 304 stainless steel (maximum carbon content of 0.05
           percent) with a minimum average outside diameter of
           5/8 inch. Shells shall be carbon steel.

4.2.2.1   For BWR units, each heater shall be a closed, vertical, channel
           down, U-tube type with provisions for shell pulling to
           provide access to tube bundles for maintenance. Tubes
           shall be Type 304 stainless steel (maximum carbon content
           of 0.05 percent) with a minimum average outside diameter
           of 5/8 inch. Shells shall be carbon steel. Cobalt content
           shall not exceed 0.05 percent by weight in stainless steel
           and 0.01 percent by weight in carbon steel.

4.2.2.2   The heads or water channels of the heaters shall be of the
           hemispherical or elliptical type, all welded construction
           with integral tube sheet.

4.2.2.3   Velocity in the tubes at rated flow and 60°F shall not exceed
           10 feet per second.

4.2.2.4   Channel relief valves shall be included on each No. 1
           feedwater heater.

4.2.2.5   The design pressure of the channel side of the feedwater
           heaters shall be the same as the main feed pump design
           pressure (see section 4.2.1.6).

4.2.2.6   Required steam for feedwater heating shall be furnished
           by the extraction system. Feedwater heaters shall include
           nozzles for extraction, drain, and vent lines as required.

4.2.2.7   Manways (minimum 20-inch I.D.) shall be provided for access into the channels of the heaters for inspection and maintenance of the tube sheets and plugging of leaking tubes.

4.2.3   Piping

4.2.3.1   All piping in the feedwater system upstream of the outermost seismic anchor shall be designed in accordance with ANSI B31.1. Design pressure of piping and valves shall be equal to that given for the feed pumps (see section 4.2.1.6) divided by 1.2 (20 percent over stress allowance permissible under ANSI B31.1).

4.2.3.2   Flow elements supplied by the NSSS contractor shall be installed in each main feedwater line to the NSSS downstream of No. 1 heater. These flow elements provide the feedwater flow signal required by the feedwater control system. In addition, the output of the flow element shall be transmitted to the control room for feedwater flow indication.

4.2.3.3   The cleanup recirculation system line shall be sized to provide a minimum of 33-1/3 percent feedwater system design flow during startup operations with the hotwell and booster pumps operating.

4.2.3.4   Provisions shall be made for insertion of a calibrated flow section downstream of the No. 1 feedwater heater for use during the turbine acceptance test. This shall include an adequate length of straight pipe for the proper accuracy of the flow element. At the conclusion of the turbine acceptance test, the test section will be removed and replaced by a straight pipe section.

4.2.3.5   Piping should be arranged to allow temporary connections for chemical cleaning of the main feedwater piping up to, but not including, the pumps or feedwater regulating valves.

4.2.3.6   The main feedwater piping should be arranged, supported, and protected as required to prevent multiple accidents in the case of rupture of a main feedwater or other (main steam) pipe breaks. The effect of rupture of feedwater lines on the containment check valves should be determined (i.e., a water-hammer analysis should be performed).

4.2.3.7   Piping design shall include considerations to ensure against excessive reactions on the pumps, feedwater heaters (limits to be recommended by the manufacturers), and the NSSS interface. The piping shall include provisions for draining all low points in the system. Piping should be arranged to slope up to the NSSS feedwater inlets so as to minimize the possibility of water hammer.

4.2.3.8   Pipe sizes shall be chosen to limit water velocity to approximately 20 fps at 100 percent load with the unit in normal operation configuration.

4.2.4   Main Feed Pump Bypass

4.2.4.1   For PWR units, the main feed pump bypass shall connect the feed pump suction header to the feed pump discharge header and shall contain a check valve followed by a motor-operated MFP bypass valve that can be manually operated from either a local panel or the control room. The MFP bypass valve shall be provided with limit switches for control room indication of valve open and closed position. Once this valve begins to close, it shall not be permitted to stop or reopen until the valve has closed completely.

For BWR units, the main feed pump bypass shall connect the feed pump suction header to the feed pump discharge header and shall include the following components:

a.   A feed pump bypass valve (FPBV) in series with a check valve to control flow of condensate from the feed pump suction header to the inlet of the startup valve station.

b.   A startup valve station comprised of the startup level control valve (SLCV) and startup level control bypass valve (SLCBV).

c.   Piping as required with provisions for connecting lines from each feed pump isolation bypass valve to the inlet of the startup valve station.

The FPBV and SLCBV shall be motor-operated with provisions for manual control from a local panel and manual or automatic control from the main control room. The FPBV and SLCBV shall be provided with limit switches for control room indication of valve open and closed position. Once these valves begin to close, they shall not be permitted to stop or reopen until the valve has closed completely.

4.2.5   Feedwater Control Valves (PWR Only)

A feedwater control valve and an auxiliary control valve supplied by the NSSS contractor control the flow of feedwater to each steam generator. The valves are air-operated and are controlled by the feedwater control system. Each feedwater control valve and auxiliary control valve can be controlled by the unit operator in the main control room using the feedwater control system "master control station."

4.2.6   Startup Level Control (BWR Only)

4.2.6.1 Startup Level Control Valves (SLCV)

A low flow bypass control valve shall be provided for startup level control. The startup level control valve can be controlled by the unit operator in the main control room using the feedwater control system "master control station," or automatically by the single element level controller (from the NSS-supplied feedwater control system).

4.2.6.2 Isolation Valve Bypass Valves

A connection shall be provided in the discharge piping of each MFP (between the positive closing check valve and the isolation valve) to permit diverting feedwater to the inlet of the start-up valve station through individual (one per pump) isolation valve bypass valves (IVBV). The IVBV shall be motor-operated, controlled by manual or automatic signal from the main control room, and equipped with limit switches for control room indicatio of valve open or closed position. Logic shall be provided to prevent valve motion from stopping in other than fully open or closed position.

4.2.7   Positive Closing Check Valves

Each main feed pump shall be equipped with a positive closing check valve (PCCV) in the discharge piping for pump protection. Each PCCV shall be equipped with a limit switch to provide control room indication if the valve moves away from the fully open position.

4.2.8   Isolation Valves

The suction and discharge lines from each main feed pump, the inlet lines to the No. 2 heaters, and the discharge lines from the No. 1 heaters shall be equipped with motor-operated, isola-tion valves. These valves shall be provided with limit switches for control room indication of valve position and shall be capable of manual actuation from either the control room or local panels. Once these valves begin to close they shall not be permitted to stop or reopen until the valve has closed completely.

4.2.9   Cleanup Recirculation Valves

4.2.9.1 To accommodate the startup requirements of section 3.1, a connection shall be provided in the full flow section downstream of the high pressure heaters to recirculate

feedwater back to the main condenser through cleanup recirculation valves in series. These valves shall be sized to pass 1/3 feedwater system design flow with only hotwell and condensate booster pumps operating. (Two valves in series are required to prevent leakage during normal operation.)

4.2.9.2   The cleanup recirculation valve shall be motor operated and shall be capable of remote manual positioning from the control room or a local panel. Limit switches shall be provided for control room and local panel indication of valve open and closed position.

4.2.10   Instrumentation

4.2.10.1   In addition to the instrumentation previously described or required to perform the logical functions described herein, additional instrumentation shall be provided to measure system temperatures and pressures required to satisfy the input needs of the balance of plant computer system, provide test information as required under ASME PTC-6, and give additional information regarding operation of the system equipment. As a minimum, this shall include:

1.   Temperature measurements at the suction of each pump and the inlet and outlet of each heat exchanger.

2.   Pressure measurements at the suction and discharge of each pump.

3.   Pressure measurements at inlet and outlet of each heat exchanger or stream of feedwater heaters.

If the output of the sensing elements is not required for plant computer or logic functions, output display on local panels shall be sufficient.

4.2.10.2   Means shall be provided to alert the operator of a low net positive suction head (NPSH) condition at any of the MFP's. These signals shall be developed by comparing individual pump suction pressure to No. 3 heater shell average pressure. "Low NPSH" and "low-low NPSH" shall be alarmed when differential pressure falls below predetermined values (see section 4.3.4.5). The alarm indicators will be in the control room. An automatic unit load runback will be initiated based on "low-low NPSH."

4.2.10.3   Vibration detectors shall be installed on the main feed pump to record and monitor shaft vibration. Alarms in the main control room shall be set to alarm on high vibration of approximately 2.0 mils double amplitude.

4.2.11 Feedwater Control System

4.2.11.1 Westinghouse and Combustion Engineering NSSS Only--The
condensate and main feedwater system is designed to
automatically maintain the steam generator water level
during steady state and transient conditions with NSSS
power of 15 percent to 105 percent. The feedwater control
system supplied by the NSSS contractor uses independent
parallel control systems, one per steam generator to
maintain the water level within acceptable limits by
controlling the feedwater control valves and the main
feed pump turbine speed setpoint. The feedwater control
system is a three-element control system using feedwater
flow, steam flow, and steam generator water level as
inputs for automatic level control above 15 percent NSSS
power. The output of each three-element system provides
signals to position the respective feedwater control
valves. In addition, the feedwater control system
simultaneously provides a pump speed setpoint signal based
on flow requirements of the steam generators.

Babcock and Wilcox--The condensate and feedwater system
is designed to automatically maintain the necessary steam
generator flow during steady state and transient conditions
with NSSS power of 15 percent to 105 percent. The feedwater
control system supplied by the NSSS contractor is an
integrated control system using megawatt demand and steam
pressure (as long as system frequency is maintained
within limits) to position the feedwater control valves
and pump speed setpoint.

General Electric BWR--The condensate and feedwater system
is designed to automatically maintain reactor vessel water
level during steady state and transient conditions with NSSS
power of 0 percent to 105 percent. The feedwater control
system supplied by the NSSS contractor maintains reactor
vessel water level within acceptable limits by controlling
the main feed pump turbine speed setpoint and/or the SLCV
position. The feedwater control system utilizes both a
single element (reactor vessel level) controller which
can provide input to both the turbine speed or the SLCV
and a three-element control system using feedwater flow,
steam flow, and reactor vessel water level as inputs for
automatic control above 15 percent NSSS power by changing
turbine speed.

4.2.11.2 The feedwater control system may be operated in manual
control mode by the operator at any power level. When in

33

manual control, the operator can use a "master control station" in the control room to adjust feedwater flow.

4.2.11.3   For BWR units, a bias signal is added to the single element control signal going to the SLCV to force the system to control reactor level at a higher setpoint when utilizing the SLCV in order to permit a completely automated startup of the main feed pumps.

4.2.12   Insulation

All piping and equipment whose temperature may exceed 150°F shall be insulated to provide surface temperatures that do not exceed 125°F with ambient air at 80°F and 50 ft/min velocity.

4.3   LOGIC FOR OPERATION

4.3.1   Main Feed Pump Startup

4.3.1.1   The unit startup will be done with one MFP in addition to the two hotwell pumps (HWP) and two condensate booster pumps (CBP) (assuming the condensate system is started automatically). The HWP's and CBP's are to be operating on their recirculating control as described in the Design Guide DG-M2.3.1, "TURBINE GENERATOR HEAT CYCLE--Condensate System--Nuclear." The following interlocks must be satisfied before the MFP turbine can be start

a.   The lube oil system for the MFP and turbine must be operating with required lube oil pressure.

b.   The MFP minimum flow recirculating control valve must be open. (Recirculating control valve can be opened only after lube oil pressure is satisfied.)

c.   The raw water isolation valves (inlet and outlet) for the corresponding MFP turbine condenser must be open.

d.   The MFP suction valve must be open, and the required MFP suction pressure must be available for pump operation. (The required MFP suction pressure should be equal to minimum required suction pressure at 100 percent unit load.) The suction pressure permissive required to start the MFP will <u>not</u> trip the pump after it is operating.

4.3.1.2   Normal procedure for warmup of the second and third MFP will be to open the pump suction valve for approximately 1 - 3 minutes (final value to be approved by pump manufacturer) prior to starting the respective feed pump turbine. Since

34

the recirculation valves are normally open when the MFP is not operating, feedwater will then flow through the pump and recirculation valves and back to the condenser.

If the second and third MFP are isolated from the feedwater circuit during startup, the warmup rate recommended by the pump manufacturer (or 100°F rise per hour if unspecified by the manufacturer) should be followed. This warmup will be done with the MFP recirculating control valve.

4.3.2   Main Feed Pump Lube Oil System

4.3.2.1 The main feed pump turbine turning gear shall be interlocked against starting unless pump bearing oil pressure is sufficient. Pressure switches shall be arranged to alarm on low bearing oil pressure and to trip the main feed pump and turbine on a low-low bearing oil pressure. The pump suction valves shall be interlocked against opening unless the pump and turbine bearing oil pressures are sufficient. Bearing oil and other protective trips shall be independent (for each pump).

4.3.2.2 The loss of lube oil pressure will automatically close the MFP recirculating control valve after a one-minute delay, to prevent rotation of the turbine and pump due to water recirculating flow through the MFP.

4.3.3   Main Feed Pump Valves

4.3.3.1 Each MFP has its own minimum recirculating control valve that will open when the minimum flow requirements, metered in each MFP suction, falls below the minimum design flow for the MFP. Minimum design flow should be specified by the MFP contractor. The recirculating control valve will modulate closed when feedwater flow through the MFP exceeds the minimum flow requirements. These valves will be designed to fail open.

4.3.3.2 The initial startup of the first MFP (at approximately 600 psi) will be done with a closed MFP discharge valve. The MFP will start up on minimum recirculating flow control.

4.3.3.3 PWR Only--After the MFP turbine is brought up to a minimum speed (specified by turbine and pump manufacturer) and stabilized the MFP discharge valve will be opened by the operator to supply feedwater through the auxiliary feedwater regulating valve as required by the steam generator throughout its pressurization to no-load pressure.

4.3.3.4 BWR Only--After the MFP is brought up to a minimum speed (specified by turbine and pump manufacturer) and stabilized, the MFP bypass isolation valve will be opened by the operator to supply feedwater through the startup level control valve as required by the reactor throughout its pressurization.

4.3.3.5   PWR Only--The feedwater bypass circuit around the MFP's shall be opened for recirculating mode cleanup of feedwater piping system, initial filling of the steam generator during startup, and for water requirements during pressurization of the steam generator prior to operating the first MFP.

A check valve in the MFP bypass line will prevent reverse flow when the MFP's are operating. Operation of the bypass shutoff valve shall be remote manual.

BWR Only--The feedwater bypass circuit around the MFP's shall be opened for recirculating mode cleanup of feedwater piping system, initial filling of the reactor vessel during startup and water requirements during pressurization of the reactor pressure vessel prior to operating the first MFP, and filling the containment pool for refueling operations.

4.3.4   Main Feed Pump Trip and Runback

4.3.4.1   In the event of a trip of one MFP, when three main feed pumps are in operation, the feedwater control system will accelerate the two MFP's still operating to maintain unit load up to 100 percent of NSSS rating.

4.3.4.2   A MFP turbine trip will occur from high MFP turbine backpressure

4.3.4.3   Provisions shall be made to trip the main feed pumps from signals initiated in the NSSS (e.g., high reactor water level for BWR, feedwater isolation signal for PWR, etc.).

4.3.4.4   The shutdown or trip of a MFP will initiate a signal to close the PCCV in the discharge line of that MFP.

No automatic action is required to close the MFP suction and discharge valves. The operator may initiate closure for isolation of the MFP circuit.

4.3.4.5   Low NPSH (20 psi above the NPSH required by the MFP at 100 percent load) will be alarmed. Manual runback of unit load can be initiated to avoid loss of NPSH at the MFP suction.

With unit load greater than 65 percent, a further loss of NPSH below the alarm setpoint (10 psi above the NPSH

required by the MFP at 100 percent load) will automatically
run back unit load to 65 percent. This runback allows
for low NPSH due to 50 percent of the No. 3 heater drain
flow being bypassed to the condenser or the loss of one
No. 3 heater drain pump.

With unit load between 40 and 65 percent, a further loss of
NPSH below the alarm setpoint (10 psi above the NPSH
required by the MFP at 65 percent load) will automatically
run back unit load to 40 percent. This runback allows for
low NPSH due to the loss of two hotwell pumps, two condensate
booster pumps, or three No. 3 heater drain pumps.

The alarm or runback of unit load will occur only after a
5-second time delay to prevent load runback due to system
pressure surges, signal noise, etc.

4.3.5   Feedwater Heater Isolation

The high pressure feedwater heaters shall be arranged in three
parallel streams. Each stream shall be equipped with motor-
driven isolation valves at the inlet to the No. 2 heater and
the outlet of the No. 1 heater. The instrumentation on
the heaters shall have provisions to annunciate a high level.
In addition, further annunciation will follow at high-high
level. If after a preset time delay, the high-high level
has not subsided, the stream in which the abnormality exists
is isolated by closing the inlet and outlet feedwater valves
and the incoming steam and drain valves. It shall be required
that the stream be manually brought back on line from this type
of isolation. The isolation of a stream of high pressure
feedwater heaters (Nos. 1 and 2) results in the full flow passing
through the two remaining streams of feedwater heaters.

4.3.6   Auxiliary Control Valves

Westinghouse and Babcock & Wilcox PWR Only--Water requirements
during initial pressurization of the steam generator are
supplied by manually controlling the auxiliary control
valve. Feedwater is manually controlled using the auxiliary
control valve and main feed pump-turbine speed until the
unit reaches 15 percent rated NSSS power. When the unit
reaches 15 percent rated NSSS power, the feedwater control
is transferred to the NSSS feedwater control system which
automatically controls the feedwater control valve, auxiliary
control valve, and the main feed pump turbine speed.

Combustion Engineering PWR Only--Water requirements during initial pressurization of the steam generator are supplied by manually controlling the auxiliary control valve to the downcomer nozzle. When the NSSS system begins power operation, feedwater flow is transferred to the control valve supplying the economizer nozzles. Feedwater is manually controlled using the feedwater control valve and main feed pump-turbine speed until the unit reaches 15 percent rated NSSS power. When the unit reaches 15 percent rated NSSS power, the feedwater control is transferred to the NSSS feedwater control system which automatically controls the feedwater control valve, auxiliary control valve, and the main feed pump turbine speed.

4.4 MAINTENANCE AND INSPECTION CRITERIA

4.4.1 Design of the system shall provide for periodic functional tests and visual inspections of the system to be performed in conjunction with unit outages. Manways or removable heads shall be provided on all heat exchangers to provide access to the tube sheets for inspection, repair, or tube plugging. Design of the feedwater system shall include provisions for component isolation to permit on-line maintenance of the following equipment:

   a. Any one main feed pump

   b. Any one recirculation valve

   c. Any one stream of high pressure feedwater heaters.

The design shall include provisions for removal of the shell of any of the high pressure feedwater heaters. In addition, shielding will be provided where necessary (BWR plant only).

4.4.3 All equipment shall be equipped with drain connections to permit complete removal of water from the equipment internals.

4.4.4 BWR Only--Facilities shall be provided in the turbine building to permit decontamination of the equipment that must be returned to the factory for maintenance. For details, see DG-M12.2.1, "CHEMICAL TREATMENT--Tool and Equipment Decontamination Facilities."

4.5 SYSTEM INTERFACES

The following documents will assist in defining system interfaces and provide additional insight into the design of the main feedwater system for a 3800 megawatt-thermal nuclear plant with light water reactors:

a.   EN DES Design Guide DG-M2.3.1, "TURBINE GENERATOR HEAT CYCLE--
     Condensate System--Nuclear," Tennessee Valley Authority:
     Knoxville, (proposed 1979).

b.   EN DES Design Guide DG-M2.4.1, "TURBINE GENERATOR HEAT CYCLE--
     Extraction Steam System--Nuclear," TVA: Knoxville, (proposed
     1979).

c.   EN DES Design Guide DG-M2.5.1, "TURBINE GENERATOR HEAT CYCLE--
     Heater Drains and Vents--Nuclear," TVA: Knoxville, (proposed
     1979).

d.   EN DES Design Guide DG-M2.19.1, "TURBINE GENERATOR HEAT CYCLE--
     Main Feed Pump Turbine System--Nuclear," TVA: Knoxville,
     (proposed 1979).

APPENDIX A

DESIGN CALCULATIONS
FOR
MAIN FEEDWATER SYSTEM

1.0 GENERAL

This portion of the design guide presents the calculations necessary for the basic design of the feedwater system and for the sizing of the main feedwater pumps, prior to issuing the requisition. In addition to being an aid to the engineer responsible for the system design, it sets forth a format for these calculations. This ensures that system design is consistent among TVA plants and that the calculations are a convenient source of reference.

2.0 NECESSARY INFORMATION

The information described in sections 2.1 through 2.4 is necessary in order to perform the feedwater system calculations.

2.1 FROM NSSS CONTRACTOR

The following information should be obtained at 105 percent, 100 percent, 75 percent, 50 percent, and 25 percent of NSSS rating:

2.1.1 The final feedwater flow. For BWR units, the final feedwater flow should differ from the steam flow by the amount of flow required for control rod drive feed.

2.1.2 Steam outlet pressure and pressure drop through the NSSS from the feedwater inlet to the steam outlet.

2.1.3 For BWR units, the feedwater flow and pressure at the reactor pressure vessel inlet nozzle required for dynamic control at the maximum steady state conditions.

2.1.4 Flow required to avoid reactor scram following trip of a main feed pump during operation at 100 percent load.

2.2 FROM TURBINE GENERATOR CONTRACTOR

Turbine generator heat balances are needed for flows of 105 percent, 100 percent (same as NSSS rating), 75 percent, 50 percent, and 25 percent (Brown Boveri usually bases their heat balance flows on similar percentages of NSSS thermal rating, instead of main steam flows). These heat balances will be used primarily to obtain feedwater temperatures at various points in the cycle.

40

2.3 FROM TASK FORCE PLANT LAYOUT AND PIPING GROUP

Piping isometric sketches are needed showing lengths between changes in pipe diameter or branches in piping and approximate location of valves and fittings. The isometric should also show the relative elevations of the centerline of the main feed pump and the feedwater inlet nozzle to the NSSS. The plant layout and equipment group should refer to the Design Criteria Diagram for feedwater piping requirements. The Feedwater System Sketch is included in this design guide figure A-5 as an aid in developing the Design Criteria Diagram.

2.4 FROM MAIN FEEDPUMP CONTRACTOR (After award of contract)

Head-capacity efficiency, and brake horsepower curves are needed from pump shutoff conditions to 150 percent of pump design flow. Curves should be shown for various speeds from rated pump speed down to 10 percent of rated pump speed. Minimum recommended flow and minimum recommended speed (if any) should be shown. The NPSH requirements at various speed and flow conditions should be shown on the curve.

3.0 FEEDWATER PIPING AND FITTING PRESSURE DROP

3.1 Calculations should be performed to obtain the pressure drops in the feedwater lines at the turbogenerator guarantee point which is also 100 percent NSSS rating. The calculations should consider pipe friction loss and head losses in fittings, flow nozzles, valves, and channels of feedwater heaters. The containment isolation valves and any other valves unique to the NSSS should be included in these calculations. For PWR units, the pressure drop in the feedwater control valve should not be considered in these calculations. Methods described in DG-M2.11, "Pressure Drop Calculations for Steam and Condensate Piping and Fittings," should be used in making these calculations.

3.2 Pipe sizes should be chosen to limit feedwater velocity at 100 percent T-G load to a maximum of 20 ft/sec. Prior to the purchase of the main feed pump, schedule 120 pipe should be assumed for sizes 10 inches and larger.

3.3 The piping isometric sketches from the Task Force Plant Layout and Piping Group should be used to obtain piping lengths, number of fittings, etc.

3.4 Pressure drops caused by friction losses through sections of piping and equipment should be calculated based on the condensate required for 100 percent T-G rating, with 3 MFP's running. Pressure drops for the other T-G load cases should be calculated in direct proportion to the square of the feedwater flow:

$$H_x = H_{100} \, (Q_x/Q_{100})^2$$

H_x = Pressure drop in psi at load X where X = 25, 50, 75, 105, 115.5% NSSS rating, two-pump runout, etc.

H_{100} = Pressure drop in psi at 100% NSSS rating

Q_x = Flow at load point X in lb/hr where X = 25, 50, 75, 105, 115.5% NSSS rating or two-pump runout, etc.

Q_{100} = Flow at 100% NSSS rating in lb/hr.

Additional items that should be considered in calculating friction pressure drops in the feedwater system are as follows:

3.4.1   (BWR) The main steam pressure drop must be considered in determining pressure at the NSSS feedwater interface since the overall control of the unit requires the reactor pressure to vary as required to maintain desired pressure at the inlet of the T-G stop and control valves.

3.4.2   Each PWR NSSS contractor has a different method of determining the pressure drop across the feedwater control valve, and should be requested to furnish this pressure drop value at the load points required for this calculation. For preliminary estimates of the feedwater control valve pressure drop, the following criteria can be used:

3.4.2.1   Westinghouse--The feedwater control valve pressure drop at each load point should be equal to the system friction loss from the MFP discharge header to the main steam header.

3.4.2.2   B&W--The feedwater control valve pressure drop should be 50 psi, constant for all loads.

3.4.2.3   Combustion Engineering--The feedwater control valve pressure drop should be 40 psi, constant for all loads.

3.5   In calculations of static head in this design guide, it can be assumed that the variation in feedwater temperature with load has a negligible effect on static head, and the static head at 100% T-G load can be used at all loads. Pressure can be converted from feet of head to psi as follows:

psi = (feet static head)/(144 x SV)

                or

psi = (feet static head) x (density)

where SV = specific volume in ft^3/lb_m and density is expressed in lb /ft-in^2.

42

3.6 The pump head available is normally given by vendors in feet of water. This can be converted from feet of water to psi, as needed, when using tables A-1, and A-2.

3.7 For calculation of MFP head (H), the total friction pressure drop and the total static pressure drop in the feedwater system (PWR) or feedwater system and steam systems (BWR) should be tabulated.

4.0 SIZING CALCULATIONS FOR MAIN FEED PUMP

4.1 CRITERIA FOR MAIN FEEDWATER PUMPS (MFP) COMPONENT SIZING

4.1.1 Based on TVA's experience with MFP's for both PWR and BWR light water nuclear reactors, the most critical sizing parameter for the MFP's occurs with 2 MFP's at 100 percent T-G load with all heater streams in service and maximum demineralizer pressure drop.

4.1.2 The MFP suction pressure for all design points and partial load operating conditions shall be taken from the calculation of condensate system pressures as determined in Appendix B of DG-M2.3.1 "Condensate System--Nuclear." The MFP suction pressure from these calculations is sufficient to provide, at least, the minimum NPSH of 300 ft for the MFP rated condition and 200 ft for the MFP best efficiency point.

4.1.3 The combined pressure loss for both No. 1 and 2 heaters, with all three streams operating, shall be a maximum of 25 psi with flow corresponding to 100 percent T-G load.

4.2 CALCULATION OF "HEAD REQUIRED" CURVES

The total feedwater system head required (H) should be calculated using the piping isometrics and the pressure drop procedures discussed in section 3 of this appendix. The calculation should be made by following the steps given in either table A-1 or A-2. For a PWR plant, the steam generator outlet pressure and pressure drop are given vs. load by the NSSS vendor. For a BWR plant, the main steam pressure at the outermost isolation valve vs load and the reactor pressure drop are given by the NSSS vendor. The magnitude of the main steam pressure drop from the main steam isolation valve to the turbine must be supplied by TVA. A separate calculation should be made for each operating requirement to be evaluated as given in section 4.1.1 of this design guide. The head required should be portrayed graphically as shown in figure A-1.

4.3 DETERMINATION OF MFP DESIGN POINTS FOR PROCUREMENT

A curve showing head required of the MFP's as a function of total feedwater flow must be prepared for use in procurement of the MFP's (see TVA document MEB-SS-3.2, "Standard Specification for Turbine-Driven Main Feed Pumps for 3800 Megawatt-Thermal (MWT) Nuclear Plants"). The H required for the rated condition and the normal operating condition (best efficiency point), as defined in section 4.1.1.1 of this design guide and calculated using section 4.2 of this appendix, should be marked on the H required curve (see figure A-2).

5.0 CALCULATION OF EXPECTED MAIN FEED PUMP OPERATING CHARACTERISTICS

5.1 MAIN FEED PUMP RATED HEAD-CAPACITY CURVE

After researching previous MFP bids and coordinating with vendors as required, determine an "expected" MFP curve shape and rated speed which meets the MFP design points in Appendix A, section 4.3, while meeting the recommended rise to shutoff of approximately 20 to 30 percent of the head at the rated condition. The "expected" curve should be based on the maximum pressure rise from the rated condition to shutoff which is expected based on the above research and coordination (see figure A-3).

5.2 CALCULATION OF VARIABLE SPEED HEAD CURVES

The laws of dynamic similarity permit the head and capacity at any pump speed, N, to be determined for a given constant pump efficiency provided a known characteristic point exists.

$$H = H_R(N/N_R)^2$$

$$Q = H_R(N/N_R)$$

where the subscript R refers to the known, in this case rated, speed.

Thus, a pump characteristic curve for any speed can be developed. Table A-3 provides a format to facilitate this development. The pump characteristics for various speeds and for one, two, and three MFP's in operation should be plotted on the same graph as the MFP head required curves to determine expected pump operating speeds at all design points (see figure A-4).

5.3 CALCULATION OF NPSH REQUIRED AT VARIABLE SPEEDS

The NPSH required by a pump is related to the suction specific speed as follows:

$$S = \frac{NQ^{1/2}}{g(NPSH)^{3/4}}$$

44

Thus, since the suction specific speed is constant and the NPSH required at the rated condition is known, the NPSH required for a given pump at varying speeds, N, can be calculated as

$$NPSH = \left[\frac{N}{N_R} \left(\frac{Q}{Q_R} \right)^{1/2} \right]^{4/3} \times \left[NPSH_{Rated.} \right]$$

The NPSH required should be calculated based on the speed and flow at the MFP design points (as determined from figure A-4) and compared with the NPSH available.

5.4   CALCULATION OF VARIABLE SPEED EFFICIENCY

The MFP variable speed efficiency curve can be calculated using the feedwater system H required curve (Appendix A, section 4.2) and MFP variable speed H-Q curves. The variable speed efficiency should be calculated for the following conditions:

| Type of Plant | No. of Equipt. Items Operating | | | | | | |
	MFP	HWP	CBP	FW htr	Heater Drain Pumps	Demin. Press. Drop	Feedwater Design Flow
PWR/BWR	1	2	2	all	none	max.	50%
PWR/BWR	2	3	3	all	none	max.	50%
PWR/BWR	2	3	3	all	all	max.	50 to 100%
PWR/BWR	3	3	3	all	all	max.	50 to max.
PWR/BWR	3	2	2	all	all	max.	50 to 100%
PWR/BWR	3	3	3	2 streams	all	max.	50 to 100%
PWR/BWR	3	3	3	all	none	max.	100%

The feedwater flow and H required should be taken from the system head calculations. The speed and efficiency should then be calculated for the flow and H conditions. Variable speed brake horsepower can then be calculated as

$$BHP = \frac{H \times Q \times Vsp.}{3960 \times \eta_p}$$

The variable speed efficiency should be plotted vs. feedwater flow for the various conditions. This curve determines the most efficient combination of pumping equipment for a given flow condition. Table A-4 provides a format for this calculation and figure A-4 also shows an example of a variable speed efficiency curve for a 3 MFP, 3 CBP, 3 HWP, 3HDP operation.

6.0 CALCULATION OF FEEDWATER SYSTEM OPERATING PRESSURES

6.1 NORMAL

The pressure at the various points in the feedwater system may be computed as outlined in tables A-1 or A-2 (see Appendix A, section 4.2).

6.2 DESIGN PRESSURE

The MFP design pressure should be calculated based on the criteria in section 4.2.1.1 of DG-M2.2.1 and the MFP rated head-capacity curve, (Appendix A, section 5.1).

7.0 POST-AWARD CALCULATIONS

7.1 The system pressure calculations should be repeated as additional information becomes available throughout the design life of the plant. As a minimum, the calculations should be repeated at the following phases in plant design:

a. Completion of procurement of all pumps.

b. When piping layout is complete and the computer-generated isometric is available.

c. Completion of shop tests for all pumps (only if they are significantly different from the proposal curves and test curves).

d. Any time a design change is made that could result in significant changes in pressure drop calculations.

The status of the plant design should be clearly stated on the calculations cover sheet.

7.2 The system head curve should be revised any time the system pressure calculations are revised. The curve and calculation sheets should always be cross referenced.

7.3 Calculations should exist to verify that at all phases of the plant design, NPSH requirements were met and all important plant operating criteria could be satisfied.

8.0 (BWR) SIZING CRITERIA FOR STARTUP LEVEL CONTROL VALVE (SLCV)

8.1 VALVE SIZING

The SLCV must be sized to provide feedwater to the reactor at up to 500 gpm during startup prior to the point where the MFP can be switched to stable speed control with feedwater flow through the

MFP discharge isolation valves. The following operating condition should be used to size the SLCV after the MFP and MFPT operating characteristics are known:

a. (Max) 500 gpm flow through MFP bypass and SLCV with 2 HWP's and 2 CBP's operating, and a reactor pressure of 500 psia.

b. (Min) 50 gpm flow through SLCV with 2 HWP's, 2 CBP's, 1 MFP operating at low speed stop, and a reactor pressure of 500 psia.

8.2 MFP SPEED AND REACTOR PRESSURE WHEN DISCHARGE ISOLATION VALVES ARE OPENED

The reactor pressure at which the MFP discharge valve may be opened is determined by calculating the required MFP speed vs. reactor pressure for the following cases:

a. 500 gpm flow through the SLCV.

b. 500 gpm flow through the MFP discharge valve.

The reactor pressure must be high enough at switchover so that the MFP speed (with approximately 20 percent margin for undershoot) will not drop below the low-speed stop.

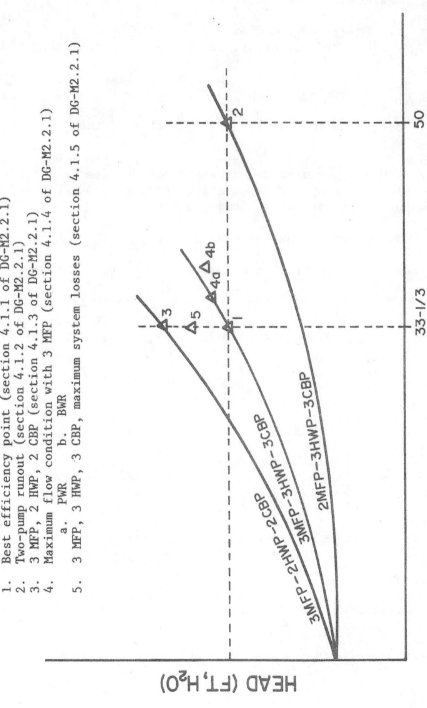

LEGEND

1. Best efficiency point (section 4.1.1 of DG-M2.2.1)
2. Two-pump runout (section 4.1.2 of DG-M2.2.1)
3. 3 MFP, 2 HWP, 2 CBP (section 4.1.3 of DG-M2.2.1)
4. Maximum flow condition with 3 MFP (section 4.1.4 of DG-M2.2.1)
   a. PWR   b. BWR
5. 3 MFP, 3 HWP, 3 CBP, maximum system losses (section 4.1.5 of DG-M2.2.1)

Figure A-1. Main Feed Pump Head Required

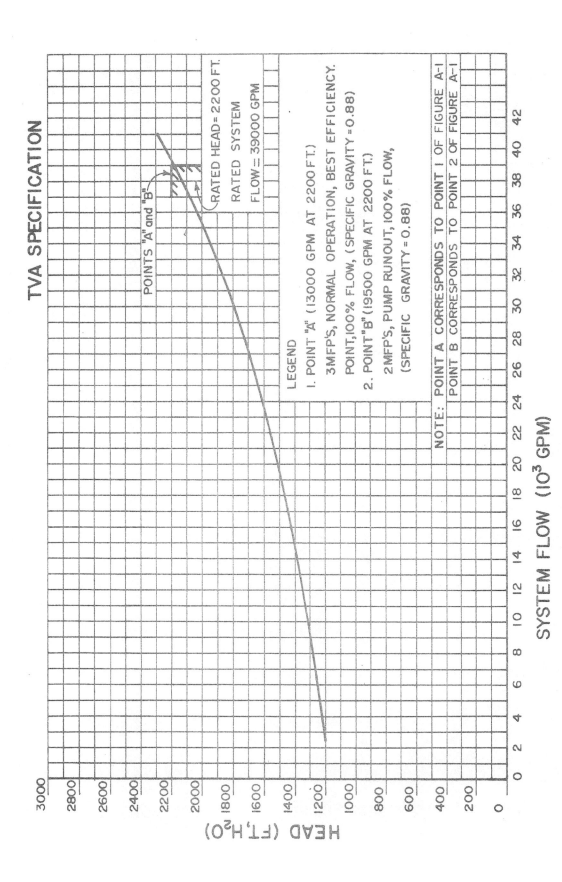

Figure A-2. Main Feed Pump Requirements (Sample)

49

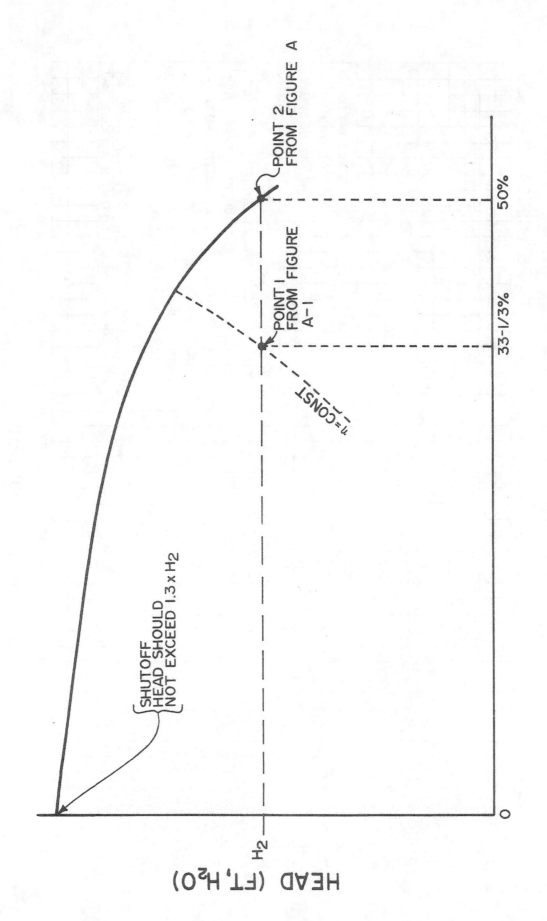

FLOW PER MFP (PERCENT TURBINE RATED FLOW)

Figure A-3. Expected MFP Characteristic

50

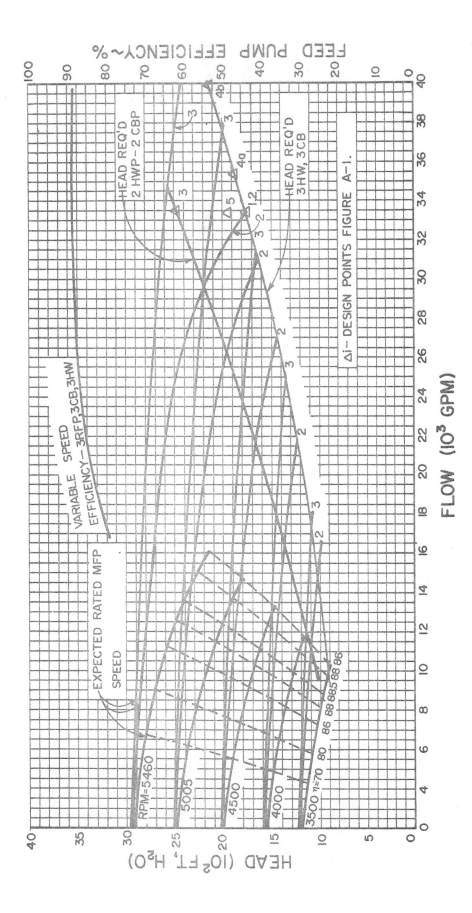

Figure A-4. Feed Pump Characteristic Curve
and System Head Requirements

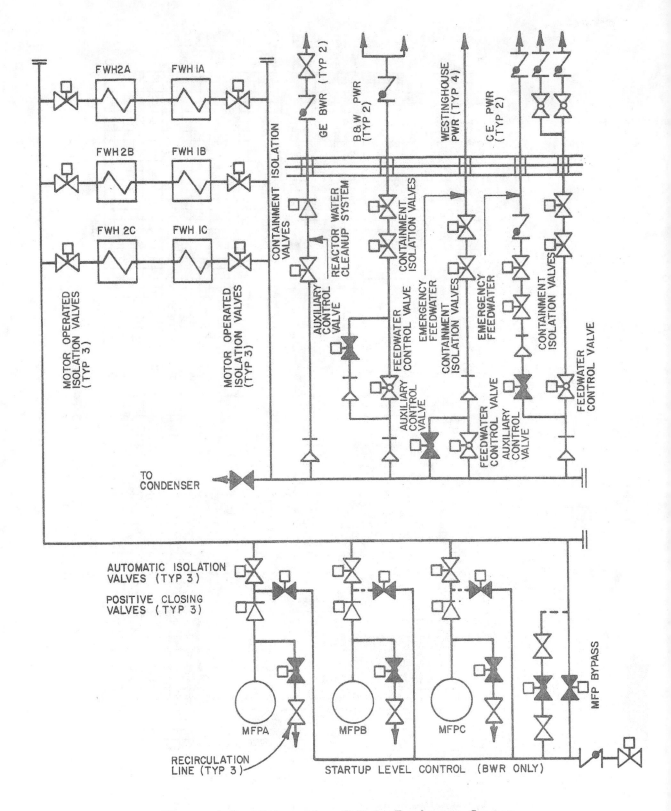

Figure A-5. Schematic of Main Feedwater System

52

TABLE A-1

Main Feed Pump Head Requirement for PWR

	Description	Units	Flow	Remarks
		% lbm/hr		
1.	Steam Generator Outlet Pressure	psia		
2.	△ P Steam Generator	psi		
3.	SG Inlet Pressure	psia		
4.	Static Head - FW Control Valve to SG inlet	ft		
5.	△P Static - FW Control Valve to SG inlet	psi		
6.	△P piping - FW Control Valve to SG inlet	psi		
7.	△P FW Flow Element	psi		
8.	Press. Downstream of FW Control Valve (3+5+6+7)	psia		
9.	△ P FW Control Valve	psi		
10.	Press at FW Control Valve inlet (8+9)	psia		
11.	Static Head - HP1 to FW Control Valve	ft		
12.	Density	$lbm/ft\text{-}in^2$		
13.	△P Static - HP1 to FW Control Valve	psi		
14.	△P Piping - HP1 to FW Control Valve	psi		
15.	Press at HP1 outlet	psia		
16.	△P - HP1	psi		
17.	Press at HP1 inlet	psia		General: Assign
18.	Static Head - HP2 to HP1	ft		positive sign to
19.	Density	$lbm/ft\text{-}in^2$		increase in static
20.	△P Static - HP2 to HP1	psi		head. (Decrease
21.	△P Piping - HP2 to HP1	psi		in elevation).
22.	△P HP2	psi		
23.	Press at HP2 inlet	psia		
24.	Static Head - MFP Disch to HP2	ft		
25.	Density	$lbm/ft\text{-}in^2$		
26.	△P Static - MFP Disch to HP2	psia		
27.	△P Piping - MFP Disch to HP2	psi		
28.	Press at MFP discharge (1+2+5+6+7+9+13+14+16+20+ 21+22+26+27)	psia		
29.	Press at MFP Suction	psia		
30.	Press required - MFP	psi		
31.	Density	$lbm/ft\text{-}in^2$		
32.	Head Required - MFP	ft		
33.	NPSH Available/Required	ft/ft		
34.	MFP Speed	rpm		

TABLE A-2

Main Feed Pump Head Requirements for BWR

	Description	Units	Flow	Remarks
		% lbm/hr		
1.	Press at main turbine control valve inlet	psia		General: Assign positive sign and
2.	△P piping - main steam isolation valve to control valve inlet	psi		increase in static head (decrease in elevation).
3.	△P NSSS (Outboard main steam isolation valve to reactor inlet feedwater sparger).	psi		0.3 psi/percent flow droop in steam press from rated press.
4.	Press at reactor inlet (1+2+3)	psia		Neglect static head of steam.
5.	Static Head - FW Flow element to reactor inlet	ft		
6.	Density	$lbm/ft-in^2$		
7.	△P Static - FW Flow element to reactor inlet	psi		
8.	△P piping - FW Flow element to reactor inlet	psi		
9.	△P - FW Flow element	psia		
10.	Press - FW Flow element inlet (4+7+8+9)	psia		
11.	Static Head - HP1 to FW Flow Element	ft		
12.	△P - HP1 to FW Flow Element	psi		
13.	△P piping - HP1 to FW Flow Element	psi		
14.	Press - HP1 outlet (10+12+13)	psia		
15.	△P - HP1	psi		
16.	Press - HP1 inlet (14+15)	psia		
17.	Static Head HP2 - HP1	ft		
18.	Density	$lbm/ft-in^2$		
19.	△P Static HP2 - HP1	psi		
20.	△P Piping HP2 - HP1	psi		
21.	△P HP2	psi		
22.	Press - HP2 inlet (16+19+20+21)	psia		
23.	Static Head - MFP inlet - HP2	ft		
24.	Density	$lbm/ft-in^2$		
25.	△P Static - MFP inlet - HP2	psi		
26.	△P Piping - MFP inlet - HP2	psi		
27.	Press at MFP Disch. (22+25+26)	psia		
28.	Press at MFP Suction	psia		
29.	Press required - MFP (27+28)	psia		
30.	Density	$lbm/ft-in^2$		
31.	Head required - MFP	ft		
32.	NPSH Available/Required	ft/ft		
33.	MFP Speed	rpm		

54

Chapter 3
Design Guide DG-M2.3.1: Condensate System

1.0 <u>GENERAL</u>

   1.1 SCOPE

   This design guide is an aid for the design of the non-safety-related
   condensate system of a 3800 megawatt-thermal nuclear plant with a
   light-water reactor.

   Use of this design guide permits preparation of a draft design
   criteria for a specific project. An engineer with a thorough
   understanding of the system should prepare the design criteria and
   perform the design calculations. Appendix A shows sketches of a
   typical system arrangement of both boiling water reactor (BWR) and
   pressurized water reactor (PWR) plants. Appendix B provides guidance
   related to system calculations for designing the condensate system
   and for establishing performance requirements needed to initiate
   procurement of the condensate system pumps.

   1.2 ABBREVIATIONS

   BWR   - Boiling water reactor

   CBP   - Condensate booster pump

   CST   - Condensate storage tank

   CSTS   - Condensate storage and transfer system

   EN DES - Division of Engineering Design (TVA)

   GSC   - Gland steam condenser

   H   - Head, ft

   HEI   - Head Exchange Institute

   H_s - Net positive suction head, ft

   HWP   - Hotwell pump

   LP   - Low pressure

   LC   - Locked closed

   LO   - Locked open

   NC   - Normally closed

   NPSH   - Net positive suction head, psi

OGC   - Off-gas condenser

Pv   - Vapor pressure

PSAR   - Preliminary Safety Analysis Report

PWR   - Pressurized water reactor

SG   - Steam generator

SJAE   - Steam-jet air ejector

SV   - Specific volume

TDH   - Total discharge head

1.3 APPLICABLE CODES AND STANDARDS

All components which contain the system pressure shall be designed and constructed in accordance with the applicable requirements of the ANSI B31.1, "American National Standard Code for Pressure Piping--Power Piping," and the "ASME Boiler and Pressure Vessel Code," Section VIII, Division 1.

Feedwater heaters shall be designed and constructed in accordance with the Heat Exchange Institute's "Standard for Closed Feedwater Heaters." Heater tubes shall conform to ASTM designation A688 (Standard Specification for Welded Austenitic Stainless Steel Feedwater Heater Tubes), with allowable stress values as defined in ASTM A249 (Standard Specification for Welded Austenitic Steel Boiler, Superheater, Heat-Exchanger, and Condenser Tubes).

The system pumps shall comply with applicable sections of the Standards of the Hydraulic Institute.

2.0 FUNCTIONAL DESCRIPTION

The condensate system consists of heat exchangers, piping, valves, and pumping units for delivering water from the main condenser hotwell to the main feed pumps (MFP) suction. This delivery must be done at the flow, pressure, and temperature required to maintain adequate net positive suction head (NPSH) at the MFP during all plant operating conditions. Providing adequate NPSH permits the feedwater system to deliver water to the nuclear steam supply system (NSSS) in ample quantity and with suitable properties to satisfy the main turbine heat balance requirements and to meet all other NSSS requirements.

The condensate system also (1) delivers water directly to the feedwater system (bypassing the MFP) during startup and (2) maintains proper water inventory in the plant.

The functional requirements of the system are met by providing the equipment and components described in sections 2.1 through 2.8.

2.1   Each of three hotwell pumps (HWP) takes suction from the high pressure (HP) zone of the main condenser hotwell, and delivers water through either the gland steam condenser (GSC) in pressurized water reactor (PWR) plants or a parallel circuit composed of the GSC, off-gas condenser (OGC), and steam-jet air ejector (SJAE) in boiling water reactor (BWR) plants. A bypass with a differential pressure control valve is provided for either configuration to ensure adequate condensate flow through the heat exchanger(s) during unit startup and low load operation while restricting pressure drop and limiting flow through the heat exchanger(s) at higher load operation.

2.2   Full flow condensate polishing demineralizers are installed in the system upstream of the feedwater heaters and condensate booster pumps (CBP) to meet water quality requirements of the NSSS (see Design Guide DG-M2.9, "TURBINE GENERATOR HEAT CYCLE--Condensate Polishing Demineralizer System - Nuclear"). The condensate demineralizers influent and effluent headers are connected to a bypass control valve to automatically open on high pressure differential across the demineralizers to protect the demineralizers and prevent the loss of adequate suction pressure for the CBP's.

2.3   A condensate flow element is installed in a full flow piping section downstream of the HWP's before any other installed equipment. The flow element is designed for the maximum expected condensate flow with minimum practical pressure drop to maintain condensate flow accuracy over the range of minimum condensate flow to maximum expected flow. Instrumentation which uses a signal generated from the flow element is also provided to control the recirculation flows for the HWP's and CBP's.

2.4   Piping connections are provided to make a condensate supply available for the following:

1.   Gland seal water head tank

2.   Turbine (condenser) exhaust hood sprays

3.   Pump seal water injection systems

4.   Seal steam evaporator.

Operating pressure requirements for these systems will determine if condensate supply should be taken from the HWP or CBP discharge. A condensate supply from the CBP discharge header downstream of the demineralizers is preferred for these services.

2.5   Three parallel streams of condensate (feedwater) heaters are provided to satisfy the main turbine heat balance requirements for the unit.

Any two of the parallel streams (one stream isolated) are capable of passing the condensate flow at 100 percent guaranteed turbine heat balance load. Automatic isolating valves (upstream and downstream of heaters) are provided for groups of heaters. One group of parallel heaters is located in the condenser neck while another group is located outside of the condenser neck.

2.6 The condensate system includes a recirculating circuit designed to maintain minimum flow required for the protection of the condensate pumps during unit startup, with two HWP's and CBP's operating. The piping connection for returning the condensate to the main condenser is located downstream of the CBP's.

2.7 A bypass to condensate storage connection is provided downstream of the condensate demineralizers for removing excess condensate from the condenser hotwell to the condensate storage tanks (see Design Guide DG-M2.6.1 "TURBINE GENERATOR HEAT CYCLE--Condensate Storage and Transfer System - Nuclear"). Connections are also provided so that water from the makeup demineralizer system can be admitted directly to the main condenser hotwell for additional deaeration prior to it being diverted to the condensate storage tank through the bypass circuit.

2.8 A makeup circuit from the condensate storage tank to the main condenser maintains minimum required condensate level in the condenser hotwell for all modes of operation.

3.0 SYSTEM DESIGN REQUIREMENTS

General requirements for performance under certain plant conditions are given in this section. Proper water inventory in the turbine cycle will be maintained by the condensate system during all modes of plant operation.

3.1 NORMAL OPERATION

3.1.1 Startup

PWR only--Under startup conditions, the PWR condensate system is designed so that water can be passed from the condenser hotwell, through the demineralizers, through the bypass line around the main feed pumps, through the feedwater heaters, and back to the main condenser. This recirculating water flow is required for system cleanup to ensure that the quality of feedwater is within the limits required by the NSSS. During the cleanup operation, the system will be capable of meeting the cooling water requirements of the gland steam condenser and for cooling the steam generator blowdown. Following the cleanup operation, the system will be capable of delivering water directly to the NSSS through the feedwater control valves while bypassing the main feed pumps. When the NSSS pressure exceeds the capability of the CBP's, the first MFP will be placed in service for supplying feedwater to the steam generators. (See Design Guide DG-M2.2.1, "TURBINE GENERATOR HEAT CYCLE--Main Feedwater System - Nuclear.")

BWR only--Under startup conditions, the BWR condensate system is designed so that water can be passed from the condenser hotwell, through the demineralizers, through the bypass line around the main feed pumps, through the feedwater heaters, and back to the main condenser. This recirculating water flow is required for system cleanup to ensure that the quality of feedwater is within the limits required by the NSSS. During the cleanup operation, the system will be capable of meeting the cooling water require-ments of the gland steam condenser, steam jet air ejectors, and off-gas condenser. Following the cleanup operation, the system will be capable of delivering water directly to the NSSS through the startup level control valve while bypassing the main feed pumps. When the NSSS pressure exceeds the capability of the CBP's, the first MFP will be placed in service for supplying feedwater to the reactor. (See Design Guide DG-M2.2.1, "TURBINE GENERATOR HEAT CYCLE--Main Feedwater System - Nuclear.")

3.1.2   Shutdown

With the feedwater system available during normal unit shutdown, the condensate system will supply the feedwater required for removal of generated heat and depressurization of the NSSS. (The residual heat removal system becomes available for removing decay heat at lower NSSS pressure levels.)

The condensate system will be used to flood the reactor pressure vessel and to fill the containment pool when preparing to refuel BWR plants.

The condensate system will be used to fill the steam generator for wet layup after unit shutdown of PWR plants.

3.1.3   Hot Standby

With the unit in hot standby, the condensate and feedwater systems (with MFP's operating) will supply the required feedwater to the NSSS.

3.1.4   Power Operation from Minimum to Maximum Expected Load

During power operation the condensate system will deliver water from the condenser hotwell through the associated condensate (feedwater) heaters and related heat exchangers (gland steam condenser - PWR and BWR, steam jet air ejector -BWR, off-gas condenser - BWR) and satisfy the main feed pump suction require-ments for delivering feedwater to the NSSS for startup to maximum expected load.

3.2   TRANSIENT CONDITIONS OF OPERATION

3.2.1   The condensate pumping system design shall accommodate the design plant load change transients.

    a.   Design plant load changes for BWR:

        1.   Step load changes of ±10 percent throughout the load range of 15 to 100 percent.

        2.   Ramp load changes of ±2 percent per minute between 15 and 30 percent reactor load and ramp load changes of ±3 percent per minute throughout the load range of 30 to 100 percent.

    b.   Design plant load changes for PWR:

        1.   Step load changes of ±10 percent throughout the load range of 15 to 100 percent.

        2.   Ramp load changes of ±3 percent per minute throughout the load range of 15 to 100 percent.

3.2.2   The condensate pumping system, with three HWP's and three CBP's operating, will be capable of pumping the additional condensate flow being bypassed to the condenser with one out of three heater drain pumps out of service (about 50 percent of No. 3 drain pump flow at 100 percent load) without a reduction in load.

3.2.3   After a loss or shutdown of all main feedwater pumps, the condensate pumping system will divert to automatic recirculation control and remain in service.

3.2.4   A loss of one complete stream out of three streams of condensate (feedwater) heaters will not require a reduction in turbine load.

3.2.5   High pressure differential across the condensate demineralizers will result in automatically bypassing the condensate flow around the demineralizers to protect the equipment and prevent a reduction in load due to the capacity limitations of the condensate pumping systems.

3.2.6   Abnormally low NPSH for the main feedwater pumps as a result of malfunctions of the condensate pumping systems will result in an automatic runback of unit load (see Design Guide DG-M2.2.1, "TURBINE GENERATOR HEAT CYCLE--Main Feedwater System - Nuclear"). Low NPSH for the CBP's will result in an alarm to permit manual runback of unit load.

4.0 SPECIFIC CRITERIA

4.1 PERFORMANCE CRITERIA

4.1.1 Condensate System Pumps

In order to meet all steady state requirements and the requirements established by the transient conditions given in section 3.2, the design of the condensate system pumps shall incorporate the following design parameters:

a. With (1) two HWP, two CBP, and three MFP in service, (2) all heater streams in service, (3) pressure drop of 50 psi across the condensate demineralizers, and (4) all heater drains being pumped forward, provide approximately 115 percent of the NPSH requirement of the MFP at 100 percent load. The 15 percent margin on NPSH is specified to minimize unnecessary actuation of control room alarm.

   For plants with this operating condition, also maintain the minimum allowable pressure at the inlet to the steam seal evaporators (specified by turbine manufacturer).

b. With (1) three HWP, three CBP, and two MFP in service, (2) all heater streams in service, (3) 50 psi pressure drop across the condensate demineralizers, and (4) all heater drains being pumped forward, provide approximately 115 percent of the NPSH required by the two MFP at 100 percent load.

c. With (1) three HWP, three CBP, and three MFP in service, (2) either one complete stream of heaters isolated and 50 psi pressure drop across the condensate demineralizers or all heater streams in service and a 65 psi pressure drop across the condensate demineralizers, and (3) all heater drains being pumped forward, provide the NPSH required by the MFP at 115 percent rated feedwater flow. (Satisfying this criterion will permit accepting a step load change from full load while operating in a faulted condition.)

d. With (1) three HWP, three CBP, and three MFP in service, (2) all heater streams in service, (3) 50 psi pressure drop across the condensate demineralizers, and (4) approximately 50 percent of the rated No. 3 heater drain flow being bypassed to the condenser, provide 115 percent of the NPSH requirement of the MFP at 100 percent load. (Satisfying this criteria will permit accepting the trip of a No. 3 heater drain pump from 100 percent load without unit trip or runback.)

e. Each of the three HWP's shall be designed to take suction from the high pressure zone condenser hotwell. The HWP shutoff head shall not exceed the design pressures of 150 lb class

61

flanges and valves at the design temperature. With all heater drains being pumped forward, two HWP's shall be capable of delivering the maximum design condensate flow from the condenser hotwell to the CBP suction (including maximum ΔP of 65 psi for the condensate demineralizers) with pressure sufficient to provide about 150 percent of the NPSH required by two CBP.

f. The condensate pumping system shall have the capability of delivering water from the condenser hotwell by way of the feedwater system for filling and startup of the steam generators (PWR) or reactor vessel (BWR). A bypass circuit (including isolating valve and check valve) shall be provided at the CBP's[1] (for use when the HWP only can satisfy all requirements) and MFP's for pumping condensate to the steam generators or reactor vessel.

g. Each HWP shall be capable of delivering the required capacity (determined by the NSSS vendor) of condensate from condenser hotwell to flood the containment pool (BWR) for the refueling operation. An automated bypass valve is provided around the normal condensate makeup circuit from the condensate storage tank to the condenser for this purpose. (See Design Guide DG-M2.6.1, "TURBINE GENERATOR HEAT CYCLE--Condensate Storage and Transfer System - Nuclear.")

4.1.2 Condensate (Feedwater) Heaters

All heaters shall be designed for a nominal 5°F terminal temperature difference (TTD) and 10°F drain coolers (except heaters that have drain tanks or external drain coolers) at 100 percent guaranteed heat balance load. The heaters shall also be designed to accept 150 percent channel side flow and the resulting shell side flow without harmful vibration (unit operating at rated load with one stream of heaters out of service).

The heaters shall be specified to perform in accordance with and comply with the HEI standards for closed feedwater heaters.

The heater channel side pressure drops (based on the operating design parameters) shall be within the allowable pressure drop limitations for the complete condensate pumping system design requirements. Heater tubes shall be limited to a minimum size of 5/8-inch O.D. and water velocity is not to exceed 10 ft/sec (based on guaranteed design flow at 60°F temperature).

---

1. The CBP's of some manufacturer can be free-wheeled so that a bypass around the CBP may not be required.

4.2 COMPONENT CRITERIA

4.2.1 Condensate Hotwell Pumps and Motors

The three hotwell pumps shall be vertical, multistage, centrifugal type pumps, each design for best efficiency operation at the condensate flow corresponding to 40 percent unit load, and shall be capable of delivering the required condensate flow and head specified in section 4.1.1.

The HWP tank shall be of sufficient depth to provide adequate NPSH at all operating conditions with minimum hotwell level. Sufficient area shall be provided between the outer shell and pump parts so that there is practically no loss between the suction nozzle and the centerline of the first stage impeller.

The pump inner casing shall be cast iron or cast steel with renewable, unleaded bronze wearing rings. The first stage impeller material shall be approved by TVA after the pump contract award. Other impellers shall be unleaded bronze or chromium alloy cast steel material. Shafts shall be made of stainless steel, and shall be protected by removable sleeves.

Hotwell pumps shall have mechanical seals with an external source connection for the seal water supply from either the HWP discharge (when the pump is running) or the gland seal water system (when pump is out of service). Seal face materials shall be tungsten carbide and carbon. The pump bearings shall be water lubricated. Each HWP shall be equipped with the necessary suction and discharge vent connections for venting back to the condenser. Temporary strainers shall be installed in the suction of each HWP.

Each HWP shall be driven by a 6600-volt, 3-phase, 60-Hertz, vertical, constant speed motor. Each motor shall be equipped with six thermocouples (two per phase--one is a backup) on the windings, and one thermocouple on each bearing. Provisions shall be made for control room annunciation of the high temperature of any thermocouple.

4.2.2 Condensate Booster Pumps and Motors

The CBP's shall be single-stage, double-suction, horizontal, centrifugal type pumps, each designed for best efficiency of operation at the condensate flow corresponding to 40 percent of unit load and be capable of delivering the required condensate flow and head specified in section 4.1.1.

The condensate booster pumps shall be equipped with mechanical seals. All control valves, filters, etc., required for seal operation and seal water flow requirements shall be provided by the pump manufacturer. Piping to the seal water control valves shall be sized to restrict water velocity to 10 feet per second maximum or to restrict the piping pressure drop to the control valve to 5 psi, whichever is limiting. Each condensate booster pump will be provided with the necessary drain, vent, and cooling connections. Temporary strainers shall be installed in the suction of each CBP.

The pump casing shall be a double volute or diffuser type, made of alloy cast steel with a nominal 13 percent chromium and 4 percent nickel content. Renewable wearing rings shall be provided in the casing.

The pump impeller material shall be alloy cast steel with either a nominal 13 percent chromium and 4 percent nickel content or 11-13 percent minimum chromium alloy.

An approved type, all metal flexible coupling and coupling guard shall be furnished between the pump and motor.

Impeller wearing rings, if used, shall be of chromium alloy steel with chromium content of 11 to 14.5 percent. Case wearing rings shall be of an alloy material suitable for use with the material of the impeller or impeller wearing rings.

Each condensate booster pump shall be driven by a 6600-volt, 3-phase, 60-Hertz, horizontal, constant speed motor. Each motor shall be equipped with six thermocouples (two per phase--one is a backup) on the windings and one thermocouple on each bearing. Provisions shall be made for control room annunciation of the high temperature of any thermocouple.

The pump and motor shall be equipped with a self-contained lubricating oil system to provide forced-feed lubrication to all pump and motor bearings. The system shall include:

a. A shaft-driven oil pump.

b. A separate 230/460-volt, 3-phase, 60-Hertz, motor-driven oil pump.

c. An oil reservoir.

d. Oil coolers and filters.

e. Oil pressure controls and gages, thermometers, relief valve, oil level indicator, all oil piping and valves for a complete lube oil system, including pressure and temperature switches for automatic operation of the system.

64

4.2.3   Condensate (Feedwater) Heaters

4.2.3.1   PWR Only--Heaters shall be closed, U-tube type, horizontal heat exchangers with a minimum tube size of 5/8-inch O.D. Tubes shall be manufactured of type 304 stainless steel material and heater shells shall be carbon steel material. The heaters located outside the condenser neck shall have a provision for pulling the heater shell, and those heaters mounted in the condenser neck shall have a provision for pulling the tube bundle for access and maintenance of the heater internals. Removable flanges or manways (minimum 20 inch I.D.) are to be provided for access into the head or channel side of the heaters for inspection and maintenance of the tube sheets and plugging of leaking tubes.

BWR Only--The No. 1 and No. 2 heaters shall be vertical, channel down, closed type, U-tube heat exchangers with integral drain coolers. The low pressure heaters shall be horizontal closed type, U-tube heat exchangers. Minimum tube size shall be 5/8-inch O.D. Heater drain coolers (if separate from the heater) shall be straight tube type heat exchangers with a minimum tube size of 5/8-inch O.D. Tubes shall be manufactured of type 304 stainless steel material, with maximum cobalt content of .05 percent by weight, and heater shells shall be carbon steel material. The heaters located outside the condenser neck shall have a provision for pulling the heater shell, and those heaters mounted in the condenser neck shall have a provision for pulling the tube bundle for access and maintenance of the heater internals Removable flanges or manways (minimum 20 inch I.D.) are to be provided for access into the head or channel side of the heaters for inspection and maintenance of the tube sheets and for plugging leaking tubes.

4.2.3.2   For heaters with integral drain coolers, the velocity entering the first row of tubes of the drain cooler section shall not exceed 1.5 feet per second while operating at guaranteed flow conditions.

The design pressure for the channel side (tube side) of the heaters shall be the maximum obtainable operating pressures (HWP's and CBP's operating) of the condensate pumping system. The design pressure shall be stated in 25 psi increments. Relief valves shall be included on the channel side of the heaters (one relief valve per isolation group) to protect against overpressure due to thermal expansion of condensate when groups of heaters can be isolated.

All heaters which have a controlled level shall have shells sized for adequate water storage capacity to prevent unstable level control over the entire operating level range.

All tubes shall be eddy current tested.

Condensate nozzle connection sizes for the channel side of heaters shall conform to HEI standards and requirements. Pressure and temperature connections shall be provided on the inlet and outlet nozzle connections. Vents and drain connections shall also be provided on the channel side of the heaters.

4.2.4   Condensate Demineralizers

Full flow condensate polishing demineralization will be provided as described in Design Guide DG-M2.9, "TURBINE GENERATOR HEAT CYCLE--Condensate Polishing Demineralizer Systems." A demineralizer bypass provides protection for the elements of the deep bed demineralizer by limiting differential pressure across the bed to 50 psi. Instrumentation necessary to measure and indicate in the control room the differential pressure between the influent and effluent headers shall be provided. The bypass valve shall fail open upon loss of control air and shall be provided with limit switches for indication of opened or closed control valve and shall provide alarms when the control valve is open. High temperature protection shall be provided for the demineralizer resins when the temperature in the influent header exceeds 140°F to isolate the demineralizers. Condensate conductivity in the influent and effluent headers shall be continuously monitored.

4.2.5   Steam Generator Blowdown Drains (PWR)

Provision shall be made for the SG blowdown drains to discharge directly into the condensate system. A connection shall be provided in the influent header of the condensate demineralizers so that all blowdown drains can be processed through the demineralizers during all modes of plant operation. Protection will be provided within the steam generator blowdown system to limit condensate influent temperature to the demineralizers and to ensure design pressure of the condensate system is not exceeded. An alternate drain bypass control valve will also be provided within the steam generator blowdown system for returning the SG blowdown drains directly to the condenser. The SG blowdown flow rates will be specified by the NSSS vendor.

4.2.6   Gland Steam Condenser (BWR and PWR)

Since the equipment must be in service when the main turbine sealing steam is applied and the condenser is at vacuum, no condensate isolation valves are provided. Minimum condensate flow is assured by the differential pressure control valve bypass circuit (section 4.2.8). A flow element shall be provided to measure the condensate flow to the GSC. The gland steam condenser (GSC), flow element, and associated condensate piping shall be sized to maintain the flow recommended by the vendor with a total pressure drop less than 30 feet at all turbine loads.

4.2.7   Off-Gas Condenser (BWR)

Condensate flow through the off-gas condenser (OGC) shall be controlled by the differential pressure control valve called for in section 4.2.9. Condensate flow to the OGC shall be measured and flow indication provided. Instrumentation shall also provide for a low flow alarm in the control room. The condensate piping circuit, including the flow element for the OGC, shall be designed to pass the necessary condensate flow with the same range of pressure drop required for the GSC for all modes of plant operation. No isolating valves shall be installed for this equipment.

4.2.8   Steam Jet Air Ejector Condensers (BWR Only)

Each of the two SJAE condensers (one normally operating and one on standby) shall be equipped with automated isolation vavles on the condensate inlet and outlet. Limit switches shall be provided on the isolation valves to give control room indication of position. Initiation of the SJAE trip/automatic switchover sequence shall cause the isolation valves associated with the faulty SJAE to close after the steam supply valve is fully closed, and shall cause the isolation valves associated with the standby SJAE to open. Condensate flow indication shall be provided for the SJAE's in the control room and on a local panel. Low condensate flows shall be alarmed in the control room. The SJAE condenser, flow element, and associated condensate piping shall be sized to maintain required flow with the same range of pressure drop required for the GSC for all modes of plant operation.

4.2.9   ΔP Control Valve for GSC, (PWR and BWR) and OGC, and SJAE (BWR)

A bypass control valve shall be provided to maintain constant pressure differential (not to exceed 30 feet of water) between the condensate inlet and the condensate outlet of the parallel heat exchanger equipment (GSC, OGC, SJAE). Control room annunciation shall be provided to alarm both high and low differential pressure across this equipment. The control valve shall also be equipped with limit switches to provide control room indication of an opened or closed position. The control valve shall fail "as is" on loss of control air or loss of control signal.

4.2.10   Automatic Makeup from Condensate Storage

The control station for the automatic makeup of condensate from storage to the condenser hotwell is discussed in Design Guide M2.6.1, "TURBINE GENERATOR HEAT CYCLE--Condensate Storage and

Transfer System - Nuclear." The purpose of this automatic condensate makeup circuit is to assure adequate water level in the condenser hotwell for all modes of plant operation and to provide for condensate recirculation during initial cleanup operations.

4.2.11 Automatic Bypass to Condensate Storage

The control station for the automatic bypass from the condensate system (connected downstream of the CBP's) to storage is discussed in Design Guide DG-M2.6.1, "TURBINE GENERATOR HEAT CYCLE--Condensate Storage and Transfer System - Nuclear." The purpose of this automatic bypass circuit is to remove excess water inventory from the condenser to the condensate storage system for all modes of plant operation, and to provide for condensate recirculation to the CST during initial cleanup operations.

4.2.12 Minimum Flow Recirculation

The minimum flow recirculation circuit shall ensure adequate flow through the condensate system to satisfy the minimum flow requirements during plant startup with the two HWP's and two CBP's operating. The minimum pumping flow shall be greater than that required for thermal protection of the condensate pumps, or that specified by the pump manufacturer for continuous operation.

The condensate recirculation flow shall also provide adequate flow in the condensate cycle during plant startup with two HWP's and two CBP's for the following heat exchanger equipment:

a. Gland seal steam condenser (PWR and BWR)

b. Steam jet air ejector (BWR)

c. Off-gas condenser (BWR)

The condensate recirculating flow shall also be designed to provide at least the minimum flow required for thermal protection when one HWP is running.

The recirculation control valve shall be controlled by a signal from the main condensate flow element located downstream of the HWP's discharge header. Total condensate flow as measured by this flow element shall also be indicated in the main control room. The recirculation control valve will modulate to maintain the minimum design flow during plant operation. The valve controls shall be designed not to modulate below 20 percent stroke but to step the valve closed at that point. A loss of HWP discharge pressure, indicating a trip of the condensate pumps, will automatically close the recirculating valve to prevent vacuum conditions in the condensate cycle.

The recirculation valve shall be provided with limit switches for open and closed position indication in the control room and shall fail open on loss of air. The recirculation valve shall be isolable and have a motor-operated, positionable bypass valve. An orifice shall be installed in the bypass line, downstream of the bypass valve, sized for a 50 psi drop at the minimum recirculation flow. The bypass MOV shall be sized to pass adequate condensate flow for cleanup and flushing operations and to provide adequate cooling of the steam jet air ejector condenser and gland steam condenser with one hotwell pump running (BWR). The bypass MOV shall be equipped with limit switches for indicating open or closed position in the control room. In addition, a manual, locked-open isolation valve capable of tight shutoff shall be provided upstream of the recirculation valve to ensure no leakage to the main condenser during the acceptance test of the main turbine generator.

4.2.13 Steam Seal Evaporator Makeup (BWR)

Makeup to the steam seal evaporator shall be provided from the condensate system downstream of the No. 4 heaters. Makeup shall be automatically controlled by an LCV in the makeup supply line from the condensate system. A makeup bypass valve shall also be provided so that if the level drops below the normal control range of the makeup LCV, the makeup bypass valve shall begin to control the level. Indication that the bypass valve has left its seat shall be provided in the control room. A further decrease in level to below the operating range of the bypass valves shall result in control room annunciation. A level above the control range of the makeup LCV shall be annunciated. Both valves shall fail closed on loss of air. On-line maintenance of valves is not required. Therefore, only one motor-operated isolation valve shall be provided in the makeup line common to both LCV's. This valve shall be seal-in type with manual control provided in the main control room. The isolation valve shall automatically close on a high-high level signal from either the seal steam side or the heating steam side of the evaporator. (The heating steam and sealing steam shall also be isolated on this signal.)

4.2.14 Main Feed Pump Suction

The condensate pumping system shall supply adequate NPSH requirements to the suction of the MFP as described in section 4.1.1. Each MFP suction shall have a temporary strainer and motor operated isolating valve.

A pump bypass circuit shall be provided around the MFP. The bypass line shall include an automated isolating valve followed by a check valve and be designed to pass approximately 33 percent of the rated feedwater flow to clean up the condensate and feedwater circuits for unit startup before feedwater can be introduced to the NSSS.

The same pump bypass circuit can be used for preoperational chemical cleanup of the piping systems.

A flow element shall be installed in the suction of each MFP to control the recirculation control valve for each MFP. In addition, pressure in the MFP suction header will be measured and transmitted to the control room to generate a NPSH control signal (MFP suction header pressure minus No. 3 heater shell pressure). This signal shall be provided for the MFP alarm and runback control logic. (See nuclear main feedwater Design Guide DG-M2.2.1, "TURBINE GENERATOR HEAT CYCLE--Main Feedwater System - Nuclear.")

4.2.15 Valves

Each HWP and CBP shall have an isolation valve in the pump suction and a check valve and isolating valve in each pump discharge.

The CBP's shall be provided with a bypass circuit with a check valve and isolating valve in the line. The circuit shall be designed to pass condensate flow required by section 4.1.1.f with one HWP operating and the CBP's not in service. It may be used during unit startup and for system chemical cleaning. If the CBPs are capable of being free-wheeled without damage, no bypass circuit is required.

The condensate (feedwater) heaters shall have motor-operated isolation valves installed on the inlets and outlets of each group of heaters. The groups of heaters requiring isolation valves shall be determined by those heaters located in the condenser neck (Nos. 5, 6, and 7) and other heaters located outside the condenser neck (Nos. 4, 3).

The No. 6 and 7 heater drain system being pumped forward into the condensate system shall also have motor-operated isolating gate valves where they connect into each stream of condensate heaters. (See Design Guide DG-M2.5.1, "TURBINE GENERATOR HEAT CYCLE--Heater Drains and Vents - Nuclear.")

All condensate system valves 2-1/2 inches and larger, which are exposed to condenser vacuum conditions, shall be equipped with water seal lantern ring glands with seal water supply from the condensate head tank. All condensate system valves 2 inches and smaller, which are exposed to condenser vacuum conditions, shall be equipped with bellows seals.

4.2.16 Piping

The condensate piping system shall be designed in accordance with ANSI B31.1 (code for pressure piping). Design pressure for the hotwell pump suction shall be 50 psig/30 in. Hg vacuum. The hotwell pump discharge to the condensate booster pump suction shall be designed for the head of the hotwell pumps while operating at minimum recirculation plus the maximum HWP suction pressure. The condensate booster pump discharge to the suction of the main feedwater pumps shall have a design pressure greater than the maximum expected condensate booster pump pressure while operating at minimum recirculation flow. The piping code does allow a 20 percent increase above design pressure during abnormal conditions for a limited period of operating time. Two abnormal conditions-- (1) no vacuum in the condenser and the condenser filled with water above the tube sheet, and (2) the loss of minimum recirculation condensate flow--should be investigated to determine the maximum pressure that can be produced by the HWP's and CBP's.

A flow element shall be installed in the full flow condensate line between the HWP and the GSC to provide the flow signal required by the minimum flow recirculation system. In addition, the output of the flow element shall be transmitted to the control room for total condensate flow indication. The flow element shall be sized to produce a total differential pressure of not more than 200 inches of water at full condensate flow.

The discharge from the No. 3 heater drain pumps enters the condensate system downstream of the No. 3 heater isolation valves. The piping connection shall be located to provide adequate thermal mixing of the heater drains and the condensate before the main feed pump suction.

The piping system design shall have adequate pipe hangers, supports, and/or guides designed according to good pressure piping design. Nozzle loading of pumps, heat exchangers, and equipment (loading and stress limitations to be specified by the manufacturer) should be considered in the piping design. Good engineering practice should be followed in the design to permit draining of the complete piping system.

Some means shall be available for continuous condensate sampling at the discharge of the hotwell pumps and the demineralizer effluent header. Sampling probes shall be located in the process piping after a straight run of pipe (10 diameters where possible, with a minimum of 3 diameters) and be located in a horizontal piping run.

4.2.17 Instrumentation

In addition to the instrumentation required to perform the logical functions described herein, additional instrumentation shall be provided to measure system temperatures and pressures required to satisfy the input needs of the balance of plant computer system, provide test information as required under ASME PTC-6, and give additional information regarding operation of the system equipment. As a minimum, this shall include:

a. Temperature measurements at the suction of each pump and the inlet and outlet of each heat exchanger.

b. Pressure measurements at the suction and discharge of each pump.

c. Pressure measurements at the inlet and the outlet of each heat exchanger or stream of feedwater heaters.

If the output of the sensing elements is not required for plant computer or logic functions, output display on local panels shall be sufficient.

The main turbine performance tests will require temporary test flow elements to be installed in the condensate piping system. The preferred location is upstream of the MFP to measure total system flow. These temporary test flow sections will be supplied by the turbine manufacturer and will require consulting with them to determine where in the condensate cycle they are to be installed.

Control room indication is required for the following:

a. Current to each HWP motor and each CBP motor.

b. Pressure in the HWP discharge header, individual CBP suction lines, CBP discharge header, and the suction of each MFP.

All pressure monitoring connections shall be equipped to facilitate on-line instrument calibration without disturbing station instrumentation.

Instrumentation should be provided on the main condenser hotwell to perform the following functions:

a. Low water level will automatically open the condensate makeup control valve from condensate storage into the condenser. High water level will automatically open the condensate return to storage control valve to remove water from the cycle back to the condensate storage tanks. Abnormally high and low water level alarms and level indications will be transmitted to the control room. The quantity of water maintained in the hotwell will be sufficient to provide 4 minutes (BWR) or 2-1/2 minutes (PWR) residence time for condensate with the unit operating at 100 percent load.

b. Conductivity monitoring shall be provided in the hotwell, spaced as required, to provide early detection and location of tube leaks.

4.2.18 Insulation

All piping and equipment outside the condenser which may exceed 150°F shall be insulated so that surface temperatures that do not exceed 125°F.

4.3 LOGIC FOR OPERATION

4.3.1 The unit startup will normally be accomplished by automatically starting two HWP's from the control room. The automatic start of two CBP's will be coincident with the start of the two hotwell pumps. The third hotwell and booster pumps will be started automatically when unit load exceeds 50-60 percent. The following logic must be satisfied for the operation of the HWP's:

a. The minimum flow recirculation valve (or bypass valve) is open and a flow path exists to the minimum flow recirculation valve. Control room alarms shall be provided to assure adequate minimum flow requirements.

b. The condenser hotwell water level is above the minimum permissive level for the operation of the HWP.

c. Motor winding temperatures are confirmed to be within limits (administrative control only).

4.3.2 A condensate hotwell pump shall trip upon indication of a motor electrical fault.

4.3.3 The condensate booster pump isolation valves shall not be permitted to open until lube oil pressure required for pump protection is guaranteed.

4.3.4 The CBP auxiliary oil pumps should start on signal to open the MOV in the CBP suction, on signal to start the CBP, on a CBP trip, or on manual signals from either the unit control room or a local panel. Control room indication that the oil pumps are running shall be provided. The auxiliary oil pumps, when started by manual command, by CBP trip, or by MOV opening logic, shall continue to run until manually stopped. When the auxiliary oil pump start is initiated by the CBP start sequence, a time delay relay shall be energized on receipt of the start signal. After the time period has elapsed, the auxiliary oil pump shall trip if oil pressure is above a pressure setpoint which indicates that the shaft driven oil pumps are functioning properly. After the auxiliary oil pumps have tripped, a decrease in oil pressure to below a low pressure setpoint shall result in restarting the auxiliary pumps. After a restart of this nature, the auxiliary oil pumps shall continue to run until manually tripped.

4.3.5   The CBP controls shall permit manual startup from either a local panel or the unit control room. Capability shall also be provided to automatically start all CBP's from the turbine/BOP automatic startup program (if provided). The following permissives shall be satisfied before a CBP starts:

a.   Motor-operated isolation valves in the pump suction (requires adequate lube oil pressure) and manual valves in the discharge fully open.

b.   Sufficient NPSH is available (measurement of CBP suction pressure is sufficient to determine NPSH).

c.   Motor winding temperatures not excessively high (administrative control only).

d.   No motor electrical faults.

e.   Adequate seal water pressure exists.

A booster pump shall trip on indication of:

a.   Motor electrical fault.

b.   Low NPSH (after a suitable time delay).

c.   Low-low oil pressure (low oil pressure starts auxiliary oil pumps). Low-low oil pressure is below the setpoint for starting the auxiliary oil pump. Low-low pressure signal shall be maintained continuously for a suitable time period before the pump is tripped.

4.3.6   Automatic isolation valves shall be installed in the inlet and outlet of each group of heaters for each of the three streams of heaters. These heater isolation valves shall be automatically closed on indication of emergency high water level in the heater shell for any heater in that group. See Design Guide M2.5, Heater Drains and Vents, for operating logic.

4.3.7   The minimum flow recirculation control valve will open when the condensate flow falls below the greater of the following flows (expected to be approximately 6000 gpm for PWR's or 8000 gpm for BWR's):

a.   The minimum required condensate flow for stable pump operation with two HWP's and two CBP's operating.

b.   The minimum required condensate flow for the gland steam condenser, steam jet air ejector (BWR only), and off-gas condenser (BWR only).

74

The signal to control the minimum flow recirculation control valve shall be generated from the differential pressure across the main condensate flow element.

4.3.8   The condensate polishing demineralizers have an automatic bypass control valve to protect the equipment against high pressure differential, approximately 50 psi, and high condensate temperature, 140°F. The opening of the demineralizer bypass control valve should be alarmed in the control room. For additional criteria and operating logic, see Design Guide DG-M2.9, "TURBINE GENERATOR HEAT CYCLE--Condensate Polishing Demineralizers."

4.3.9   The operating logic for the MFP is discussed in Design Guide DG-M2.2.1, "TURBINE GENERATOR HEAT CYCLE--Main Feedwater System - Nuclear." Interface between the condensate system (feedwater pump suction) and the main feedwater system shall include the following:

a. Flow element in each pump suction to be used to control minimum recirculation flow for the feedwater pumps.

b. Motor operated isolating valves in each pump suction to be provided with interlocks for feed pump operating logic.

c. The feed pump bypass circuit (motor operated isolation valve and check valve) will be open during plant startup (before feedwater pumps are operating) for startup pressurization of the NSSS.

d. Feedwater pump operating logic shall be provided for runback of unit load and alarms from minimum NPSH (pump suction header pressure minus No. 3 heater extraction pressure) for the feedwater pumps.

4.4   MAINTENANCE AND INSPECTION CRITERIA

4.4.1   Since no component of the condensate system has been designated as having a code safety class, inservice inspection as defined in the ASME Boiler and Pressure Vessel Code, Section XI, is not required. However, design of the system shall provide for periodic functional tests and visual inspections of the system to be performed in conjunction with unit outages. Manways and removable heads shall be provided on all heat exchangers to provide access to the tube sheets for inspection, repair, or tube plugging.

Design of the condensate system shall include provisions for component isolation to permit on-line maintenance of the following equipment:

1. Hotwell pump.

2. Condensate booster pump.

3. Any one stream of feedwater heaters not located in the condenser neck.

4. Level control valves in the condensate makeup from and bypass to the condensate storage tanks.

5. Gland steam condenser exhauster blower, and controls.

6. Either SJAE (BWR only).

The design shall include provisions for removal of the tube bundles of the heaters mounted in the condenser neck and the shells of the remaining heaters, and the tube bundles of the off-gas condenser (BWR only), SJAE condensers (BWR only), and gland steam condensers. In addition, shielding will be provided where necessary in a BWR plant.

4.4.2   Design of the turbine building shall include provisions for required maintenance of all system pumps. These provisions shall include access ways to permit equipment to be moved into or out of the proper areas, lifting devices for handling large components, overhead space required to lift hotwell pump impellers, etc.

4.5.3   All equipment should be equipped with drain connections to permit complete removal of water from the equipment internals.

4.5.4   (BWR only). Facilities shall be provided in the turbine building to permit decontamination of equipment that must be returned to the factory for maintenance.

4.5   TESTING CRITERIA

4.5.1   Each heat exchanger shall receive a shop hydrostatic test at 1.5 times design pressure which shall be performed in accordance with the ASME Boiler and Pressure Vessel Code, Section VIII, Division 1, Part UG-9 (Standard Hydrostatic Test). Heat exchanger tubes shall be tested in accordance with ASME SB-111 and ASME SB-395, latest edition, including eddy current tests.

4.5.2   The operating characteristics for each pump will be established throughout its operating range by shop tests which shall be performed in accordance with the Hydraulic Institute Standards, latest edition. A hydrostatic test will be performed on the pump casings at 1.5 times the respective design pressures.

4.5.3   The condensate system shall receive a field hydrotest test at 1.5 times the respective component design pressure (where practical).

4.5.4   A complete noncritical system preoperational test, to be scoped
by EN DES, shall be performed to verify the proper functioning of
automatic components of the system whose malfunctions could produce
reactor scram or equipment damage.

4.6   <u>System Interfaces</u>

The following documents will assist in defining system interfaces
and provide additional insight into the design of a condensate
system for a 3800 MWT nuclear plant with light-water reactors.

a.   EN DES Design Guide DG-M2.2.1, "TURBINE GENERATOR HEAT CYCLE--
Main Feedwater System - Nuclear," TVA; Knoxville, TN: 1979.

b.   EN DES Design Guide DG-M2.4.1, "TURBINE GENERATOR HEAT CYCLE--
Extraction Steam System - Nuclear," TVA; Knoxville, TN: 1979.

c.   EN DES Design Guide DG-M2.5.1, "TURBINE GENERATOR HEAT CYCLE--
Heater Drains and Vents - Nuclear," TVA; Knoxville, TN: 1979.

d.   EN DES Design Guide DG-M2.7.1, "TURBINE GENERATOR HEAT CYCLE--
Condenser and Auxiliaries - Nuclear," TVA; Knoxville, TN: 1980.

e.   EN DES Design Guide DG-M2.9, "TURBINE GENERATOR HEAT CYCLE--
Condensate Polishing Demineralizers," TVA; Knoxville,
TN: 1978.

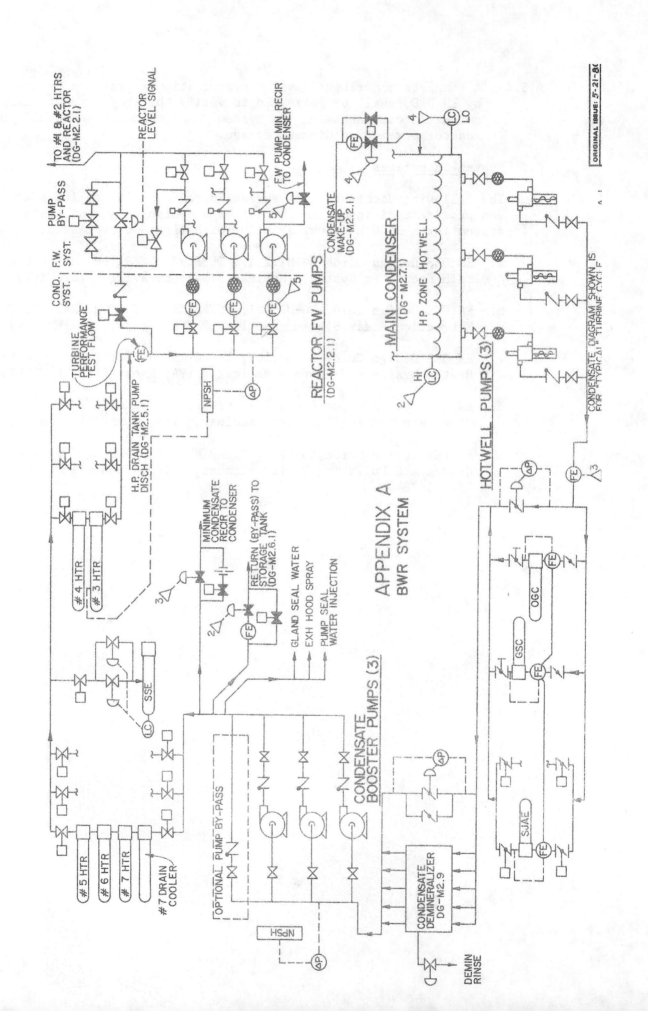

APPENDIX A
BWR SYSTEM

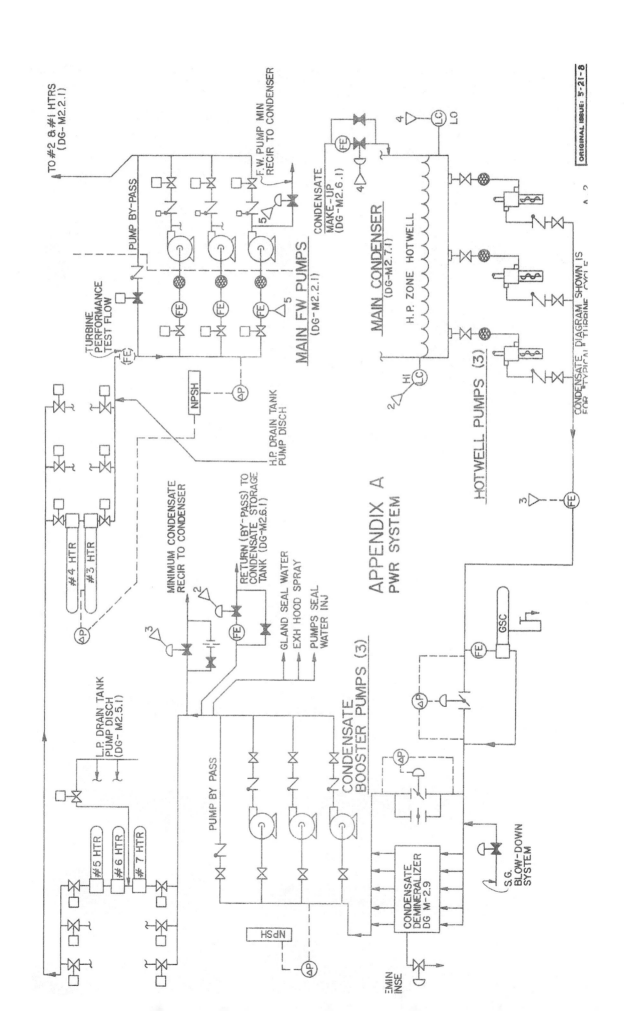

APPENDIX A
PWR SYSTEM

APPENDIX B

DESIGN CALCULATIONS
FOR
CONDENSATE SYSTEM

B.1.0 GENERAL

This portion of the design guide presents the calculations necessary for the basic design of the condensate system and for the sizing of the condensate pumps prior to the requisition. In addition to being an aid to the engineer responsible for the system design, it sets forth a format for these calculations. This ensures that system design is consistent among TVA plants and the calculations are a convenient source of reference.

B.2.0 NECESSARY INFORMATION

The information described below is necessary in order to perform the condensate system calculations.

B.2.1 FROM TURBINE GENERATOR CONTRACTOR

Turbine generator heat balances at 105 percent, 100 percent, 75 percent, 50 percent, and 25 percent of turbine rating. These heat balances will be used to obtain condensate flows and temperatures at various points in the cycle.

B.2.2 FROM TASK FORCE PLANT LAYOUT AND PIPING GROUP

Piping isometric sketches showing lengths between changes in pipe diameter or branches in piping and approximate location of valves and fittings.

B.3.0 CONDENSATE PIPING AND FITTING PRESSURE DROP

B.3.1 Design Guide DG-M2.11, "TURBINE GENERATOR HEAT CYCLE--Pressure Drop Calculations for Steam and Condensate Piping and Fittings" should be used for making calculations. The following information should be taken from the turbine-generator heat balance at 100 percent rated load:

a. Condenser pressure and condensate flow and temperature from the condensate hotwell.

b. The temperature of the condensate leaving the GSC on a PWR. The temperature of the condensate leaving the GSC, SJAE, and OGC on a BWR.

c. Condensate flow from the No. 7 heater drain pump and temperature of the condensate going to No. 6 heater on a PWR.

d. The temperature of the condensate leaving heaters No. 3, No. 4, No. 5, No. 6, and No. 7.

e. Condensate flow from No. 3 heater drain pump and temperature of the condensate going to the MFP.

f. Final feedwater flow.

B.3.2 The calculations should consider pipe friction loss and head losses in fittings, flow nozzles, and valves. Pipe sizes should be chosen to limit velocities to a maximum of 15 ft/sec. Prior to purchase of the condensate hotwell and condensate booster pumps, schedule 40 pipe should be assumed for sizes 12 inches and larger and standard schedule pipe for sizes 10 inches and smaller.

B.3.3 The piping isometric sketches from the task force plant layout and piping group should be used to obtain piping lengths, number of fittings, etc.

B.3.4 Pressure drops through sections of piping and equipment which are due to friction losses should be calculated for 100 percent turbogenerator rated flow, with three HWP and three CBPs running and all heater streams in service. The pressure drops for the other flow cases should then be calculated by multiplying the friction loss at this condition by the square of the ratio flows.

$$H_x = H_{100} \ (Q_x/Q_{100})^2$$

H_x = Pressure drop in psi at load X where X = 25, 50, 75, 105, 115.5 percent rating, 2-pump runout, etc.

H_{100} = Pressure drop in psi at 100 percent rating.

Q_x = Flow at load point X in lb/hr where X = 25, 50, 75, 105, 115.5 percent NSSS rating or 2-pump runout, etc.

Q_{100} = Flow at 100 percent NSSS rating in lb/hr.

Although the pressure drop across the gland steam condenser (PWR), or OGC, SJAE intercondenser, and gland steam condenser (BWR) is due to friction losses, the pressure drop is maintained constant by the modulating bypass control valve.

B.3.5   In calculations of static head in this design guide, it can be assumed that the variation in feedwater temperature with load has a negligible effect on static head, and that the static head at 100 percent load can be used at all loads. Although changing condensate temperature will affect static pressure can be converted from feet of head to psi as follows:

$$PSI = Feet/(144 \times SV)$$

Where SV is equal to the specific volume of water in ft^3/lb at saturation temperature of condensate at the MFP suction taken from ASME steam tables.

B.4.0   SIZING OF HOTWELL AND CONDENSATE BOOSTER PUMPS-ESTABLISHING PERMANENT SPECIFICATION REQUIREMENTS

B.4.1   INITIAL ASSUMPTIONS AND CRITERIA FOR CONDENSATE PUMP SIZING

B.4.1.1   Based on TVA's experience with MFP for all light water nuclear reactor plants, the limiting condition for sizing the condensate pumps occurs with 3 HWP, 3 CBP, and 2 MFP operating at 100 percent load with all heater streams in service and maximum demineralizer pressure drop (see section 4.1.1b of the basic design guide). All other operating requirements expressed in section 4.1.1 of the design guide will probably be met if that condition (4.1.1b) is used for pump sizing. However, all conditions described in section 4.1.1 should be checked.

B.4.1.2   The minimum NPSH required by the MFP during the runout condition (two MFPs at 100 percent of turbogenerator load) should be assumed to be 300 feet. The minimum NPSH required by MFPs at the best efficiency point (three MFPs at 100 percent of turbogenerator load) should be assumed to be 200 feet.

B.4.1.3   The total head (H) required of the condensate pumps should, as nearly as possible, be divided equally between the hotwell pumps and condensate booster pumps. However, the maximum HWP total discharge head shall be less than 255 psig (the maximum service pressure rating for 150 lb class carbon steel flanges and valves with design temp of 150°F). The CBP total head should be limited to the amount practical for a single stage, double suction pump, if possible.

B.4.1.4   The combined pressure loss for all five low pressure heater stages (including the No. 7 heater drain cooler for a BWR only), with three streams operating, should be assumed to be a maximum of 50 psi with flow corresponding to 100 percent of the turbogenerator load.

82

B.4.1.5　A maximum pressure drop of 65 psi should be assumed across the condensate polishing demineralizer. On a PWR, a pressure drop of 6.5 psi should be assumed across the gland steam condenser. On a BWR, a pressure drop of 11 psi should be assumed across the gland steam condenser, off-gas condenser, and SJAE intercondenser.

B.4.1.6　Additionally, the following unrecoverable pressure drops should be assumed for piping system components (based on 100 percent turbine rated flow with all pumps and heaters in service).

Component	Pressure drop (psi)
HWP suction strainer	2
Condensate flow element	4
MFP suction strainer	3
MFP flow element	4

B.4.2　PREPARATION OF "HEAD REQUIRED" CURVES

The head required of the condensate system pumps should be calculated by following the steps given in either table B-1 or B-2. A separate calculation should be made for each operating requirement of section 4.1.1 of the design guide to be evaluated. The head required including 5 percent margin should be portrayed graphically in a manner shown in figure B-1.

B.4.3　CALCULATION OF HOTWELL PUMP DESIGN POINT

TVA limits the maximum head rise from the design point to shutoff to 22 percent of design head in the standard specification for HWP's. Furthermore, experience has shown that maximizing the head of the HWP will generally result in the criteria of Appendix B, section B.4.1.3, being satisfied. Therefore the following procedure should be used to establish the design point for the HWP. (See figure B-2.)

B.4.3.1　Establish a preliminary design point of 206 psi $(270 \text{ psia} \times \frac{1}{1.22} \quad 15)$ at a flow that corresponds to 40 percent turbine rated.

B.4.3.2　Construct a straight line from the preliminary design point to a shutoff head of 255 psig.

B.4.3.3   After researching previous HWP bids and coordinating with
          vendors as required, determine the maximum expected "droop"
          from design point to runout and complete the HWP charac-
          teristic curve started in the preceding section B.4.3.2
          by approximating this "droop" from 40 percent to 50 percent
          turbine rated flow.

B.4.3.4   Using values for pump efficiencies determined in a similar
          manner to that used in establishing "droop" in the preceding
          section 4.3.3, compute the maximum expected motor horsepower
          required at the 50 percent flow point.

B.4.3.5   Determine what variation in head is required to reduce the
          motor horsepower to the next lowest standard size (motor
          horsepower generally varies in increments of 100 Hp below
          1000 Hp and 250 Hp and above 1000 Hp). If the head varia-
          tion required is less than 10 psi, the head at the design
          point should be reduced to permit use of the smaller motor.

B.4.3.6   Construct a family of HWP curves for one, two, and three
          pump operation superimposed on the head required curve of
          Appendix B, section B.4.2 (see figure B-3).

B.4.4   CALCULATION OF CONDENSATE BOOSTER PUMP DESIGN POINT

B.4.4.1   Determine the head required of the CBP at runout by sub-
          tracting the head available from the HWP operating at
          50 percent flow from the head required at 100 percent
          flow with three MFP operating.

B.4.4.2   Using the "worst" (steepest permissible rise from design
          point to shutoff and the quickest "droop" toward runout)
          expected CBP characteristic shape, construct a characteris-
          tic curve for the CBP (see figure B-4).

B.4.4.3   The head from this curve at 40 percent rated flow should
          be used as the CBP design point.

B.4.4.4   Perform checks to see that the pumps characteristics selected
          for the HWP and CBP meet all required performance criteria.
          Verify that NPSH requirements for the CBP are satisfied over
          the operating range (see section B.5.0). If the selected
          characteristics are satisfactory, they are sufficiently
          conservative that all pumps that meet the standard specifica-
          tion will meet all system requirements.

B.5.0 CALCULATION OF CONDENSATE SYSTEM PRESSURES

B.5.1 PREREQUISITE CALCULATION

B.5.1.1 In order to establish the head requirements of the heater drain pumps the expected pressure at various points in the condensate system must be calculated. The nature of the head required of a drain pumping system is such that the conservative approach to sizing drain pumps is to maximize the rise of the condensate system pumps between design point and shutoff. This approach was taken in section B.4.0. Therefore the system pressures calculated using the CBP and HWP characteristics established in section B.4.0 are conservative for establishing procurement parameters for the drain pumps.

B.5.1.2 The pressure at the various points in the condensate system should be computed as outlined in tables B-3 or B-4. For calculation purposes, assume each heater stage will take a 10 psi drop at 100 percent load with three streams of heaters in service.

B.5.1.3 A single stage, double suction CBP should not require more than 60 feet NPSH at the runout (50 percent rated flow) condition. The NPSH requirement at any other flow condition can be estimated as:

$$NPSH_x = 60 \text{ ft } \left(\frac{FLOW_x}{50} \right)^2$$

where $FLOW_x$ is expressed as the percentage of rated condensate flow being delivered by one CBP.

The engineer should verify that the NPSH available at the CBP (CBP suction pressure less vapor pressure less velocity head at the pump suction) exceeds the NPSH required by at least 15 percent over the entire operating range. If NPSH margin is not available alternatives include increasing HWP head (thus increasing the design pressure to above 150 lb class) or specifying multi-stage CBP.

B.5.2 POST AWARD CALCULATIONS

B.5.2.1 The system pressure calculations should be repeated as additional information becomes available throughout the design life of the plant. As a minimum, the calculations should be repeated at the following phases in plant design.

a. Completion of procurement of all pumps.

b. Completion of shop tests for all pumps.

c. Any time a design change is made that could result in significant changes in pressure drop calculations.

The status of the plant design should be clearly stated on the calculations cover sheet.

B.5.2.2 The system head curve should be revised any time the system pressure calculations are revised. The curve and calculation sheets should always be cross referenced.

B.5.2.3 Calculations should exist to verify that at all phases of the plant design, NPSH requirements were met and all important plant operating criteria could be satisfied.

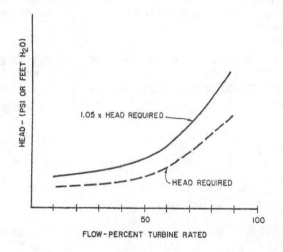

Figure B-1

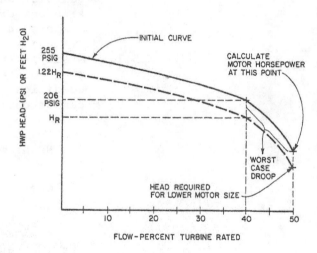

Figure B-2

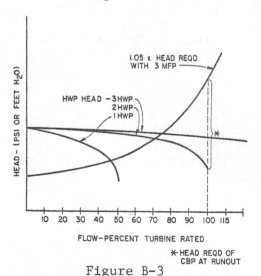

Figure B-3

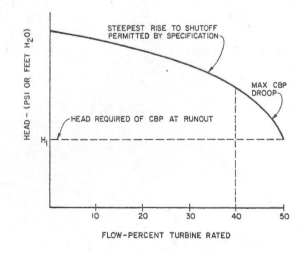

Figure B-4

86

TABLE B-1 (PWR)   SYSTEM HEAD REQUIREMENTS – NUCLEAR PLANT – CONDENSATE SYSTEM

	DESCRIPTION				UNITS			0		25	
	FLOW – COND TO NO. 7 DR. IN.				% LBM/HR						
1.	CONDENSER PRESSURE – HP ZONE				PSIA						
	STATIC HD – NWL TO HWP SUCT FLG				FT						
	DENSITY				LBM/FT IN^2						
2.	ΔP STATIC				PSI						
3.	ΔP PIPING – COND TO HWP SUCT				PSI						
	(INCLUDING SUCTION STRAINER)										
4.	HWP SUCTION PRESS (1-2-3)				PSIA						
	HEAD LOSS				FT						
	DENSITY				LBM/FT IN^2						
5.	ΔP GSC				PSI						
6.	ΔP PIPING HWP DISCH TO LOOP				PSI						
	(INCLUDING COND FLOW ELEMENT)										
	STATIC HD HWP DISCH TO LOOP				FT						
	DENSITY				LBM/FT IN^2						
7.	ΔP STATIC HWP DISCH TO LOOP				PSI						
	STATIC HD LOOP TO DEMIN INLET HDR				FT						
	DENSITY				LBM/FT IN^2						
8.	ΔP STATIC LOOP TO DEMIN INLET HDR				PSI						
	STATIC HD DEMIN INLET TO OUTLET HDR				FT						
	DENSITY				LBM/FT IN^2						
9.	ΔP STATIC				PSI						
10.	DEMIN ΔP				PSI						
11.	ΔP PIPING DEMIN DISCH TO CBP SUCT				PSI						
	STATIC HD DEMIN DISCH HDR TO CBP SUCT				FT						
	DENSITY				LBM/FT IN^2						
12.	ΔP STATIC				PSI						

TABLE B-1 (PWR)   SYSTEM HEAD REQUIREMENTS – NUCLEAR PLANT – CONDENSATE SYSTEM

	DESCRIPTION				UNITS						
	FLOW – COND TO NO. 7 DR. IN.				% LBM/HR				0		25
13.	ΔP PIPING – CBP to LP7				PSI						
	STATIC HD – CBP to LP7				FT						
	DENSITY				LBM/FT IN^2						
14.	ΔP STATIC				PSI						
15.	ΔP PIPING LP7–LP6				PSI						
	STATIC HD LP7–LP6				FT						
	DENSITY				LBM/FT IN^2						
16.	ΔP STATIC LP7–LP6				PSI						
17.	ΔP PIPING LP6–LP5				PSI						
	STATIC HD – LP6–LP5				FT						
	DENSITY				LBM/FT IN^2						
18.	ΔP STATIC – LP6–LP5				PSI						
19.	ΔP PIPING LP5–LP4				PSI						
	STATIC HD – LP5–LP4				FT						
	DENSITY				LBM/FT IN^2						
20.	ΔP STATIC				PSI						
21.	ΔP PIPING LP4–LP3				PSI						
	STATIC HD LP4–LP3				FT						
	DENSITY				LBM/FT IN^2						
22.	ΔP STATIC				PSI						
23.	ΔP PIPING LP3 TO NO. 3 DR. IN.				PSI						
	STATIC HD LP3 TO NO. 3 DR. IN.				FT						
	DENSITY				LBM/FT IN^2						
24.	ΔP STATIC				PSI						

88

TABLE B-1 (PWR)   SYSTEM HEAD REQUIREMENTS - NUCLEAR PLANT - CONDENSATE SYSTEM

	DESCRIPTION			UNITS					
	FLOW - COND TO NO. 7 DR. IN.			%LBM/HR				0	25
25.	ΔP PIPING - NO. 3 DR. IN TO MFP SUCT			PSI					
	(INCLUDING SUCTION STRAINED)								
	STATIC HD - NO. 3 DR. IN TO MFP SUCT			FT					
	DENSITY			LBM/FT IN^2					
26.	ΔP STATIC			PSI					
27.	ΔP FLOW ELEMENT IN MFP SUCTION			PSI					
28.	ΔP LP HEATERS 3-7			PSI					
29.	MFP NPSH REQUIRED			FT					
30.	VAPOR PRESSURE AT MFP SUCTION			PSIA					
31.	DENSITY			LBM/FT IN^2					
32.	MFP NPSH REQUIRED (29 x 31)			PSI					
33.	HEAD REQUIRED OF COND PUMPING								
	UNITS (2+3+5+6+7+8+9+10+11+12								
	+13+14+15+16+17+18+19+20+21+22								
	+23+24+25+26+27+28+30+32-1)								
34.	TOTAL HEAD REQUIRED INCLUDING			PSI					
	5% MARGIN (33 x 1.05)								

TABLE B-2 (BWR)   SYSTEM HEAD REQUIREMENTS – NUCLEAR PLANT – CONDENSATE SYSTEM

	DESCRIPTION				UNITS						0	25		
	FLOW – COND TO NO. 3 DR. IN.				% LBM/HR									
1.	CONDENSER PRESSURE – HP ZONE				PSIA									
	STATIC HD – NWL TO HWP SUCT FLG				FT									
	DENSITY				LBM/FT IN^2									
2.	ΔP STATIC				PSI									
3.	ΔP PIPING COND TO HWP SUCT				PSI									
	(INCLUDING SUCTION STRAINER)													
4.	HWP SUCTION PRESS (1-2-3)				PSIA									
	HEAD LOSS SJAE, OGC, GSC				FT									
	DENSITY				LBM/FT IN^2									
5.	ΔP SJAE, OGC, GSC				PSI									
6.	ΔP PIPING HWP DISCH TO SJAE, OGC,				PSI									
	GSC LOOP (INCLUDING COND FLOW ELEMENT)													
	STATIC HD HWP DISCH TO LOOP				FT									
	DENSITY				LBM/FT IN^2									
7.	ΔP STATIC HWP DISCH TO LOOP				PSI									
	STATIC HD LOOP TO DEMIN INLET HDR				FT									
	DENSITY				LBM/FT IN^2									
8.	ΔP STATIC LOOP TO DEMIN INLET HDR				PSI									
	STATIC HD DEMIN INLET TO OUTLET HDR				FT									
	DENSITY				LBM/FT IN^2									
9.	ΔP STATIC				PSI									
10.	DEMIN ΔP				PSI									
11.	ΔP PIPING DEMIN DISCH TO CBP SUCT				PSI									
	STATIC HD DEMIN DISCH HDR TO CBP SUCT				FT									
	DENSITY				LBM/FT IN^2									
12.	ΔP STATIC				PSI									

TABLE B-2 (BWR)   SYSTEM HEAD REQUIREMENTS – NUCLEAR PLANT – CONDENSATE SYSTEM

	DESCRIPTION				UNITS				0	25	
	FLOW – COND TO NO. 3 DR. IN.				% LBM/HR						
13.	ΔP PIPING – CBP TO DC7				PSI						
	STATIC HD – CBP TO DC7				FT						
	DENSITY				LBM/FT IN^2						
14.	ΔP STATIC				PSI						
15.	ΔP PIPING DC7 TO LP7				PSI						
	STATIC HD – DC7 TO LP7				FT						
	DENSITY				LBM/FT IN^2						
16.	ΔP STATIC				PSI						
17.	ΔP PIPING LP7–LP6				PSI						
	STATIC HD – LP7–LP6				FT						
	DENSITY				LBM/FT IN^2						
18.	ΔP STATIC				PSI						
19.	ΔP PIPING LP6–LP5				PSI						
	STATIC HD – LP6–LP5				FT						
	DENSITY				LBM/FT IN^2						
20.	ΔP STATIC – LP6–LP5				PSI						
21.	ΔP PIPING LP5–LP4				PSI						
	STATIC HD – LP5–LP4				FT						
	DENSITY				LBM/FT IN^2						
22.	ΔP STATIC				PSI						
23.	ΔP PIPING LP4–LP3				PSI						
	STATIC HD – LP4–LP3				FT						
	DENSITY				LBM/FT IN^2						
24.	ΔP STATIC				PSI						

TABLE B-2 (BWR)   SYSTEM HEAD REQUIREMENTS – NUCLEAR PLANT – CONDENSATE SYSTEM

	DESCRIPTION				UNITS					0		25	
	FLOW – COND TO NO. 3 DR. IN.				% LBM/HR								
25.	ΔP PIPING LP3 TO NO. 3 DR. IN.				PSI								
	STATIC HD LP3 TO NO. 3 DR. IN.				FT								
	DENSITY				LBM/FT IN^2								
26.	ΔP STATIC				PSI								
27.	ΔP PIPING – NO. 3 DR. IN. TO MFP SUCT				PSI								
	(INCLUDING SUCTION STRAINER)												
	STATIC HD – NO. 3 DR. IN. TO MFP SUCT				FT								
	DENSITY				LBM/FT IN^2								
28.	ΔP STATIC				PSI								
29.	ΔP FLOW ELEMENT IN MFP SUCTION				PSI								
30.	ΔP LP HEATERS 3-7 (INCLUDING 7DC)				PSI								
31.	MFP NPSH REQUIRED				FT								
32.	VAPOR PRESSURE AT MFP SUCTION				PSIA								
33.	DENSITY				LBM/FT IN^2								
34.	MFP NPSH REQUIRED (31 x 33)				PSI								
35.	HEAD REQ'D OF COND PUMPING UNITS				PSI								
	(2+3+5+6+7+8+9+10+11+12+13+14+15+												
	16+17+18+19+20+21+22+23+24+25+26+												
	+27+28+29+30+32+34-1)												
36.	TOTAL HEAD REQUIRED INCLUDING 5%				PSI								
	MARGIN (35 x 1.05)												

TABLE B-3 (PWR)   CALCULATIONS OF ABSOLUTE PRESSURE - NUCLEAR PLANT - CONDENSATE SYSTEM

	DESCRIPTION FLOW - COND TO NO. 7 DR. IN.				UNITS % LBM/HR			0		25	
1.	CONDENSER PRESSURE - HP ZONE				PSIA						
	STATIC HD - NWL TO HWP SUCT FLG				FT						
	DENSITY				LBM/FT IN^2						
2.	ΔP STATIC				PSI						
3.	ΔP PIPING - COND TO HWP SUCT				PSI						
	(INCLUDING SUCTION STRAINER)										
4.	HWP SUCTION PRESS (1-2-3)				PSIA						
	HWP HEAD				FT						
	DENSITY				LBM/FT IN^2						
5.	HWP PRESS RISE				PSI						
6.	HWP DISCH PRESS (4+5)				PSIA						
	HEAD LOSS, GSC				FT						
	DENSITY				LBM/FT IN^2						
7.	ΔP, GSC				PSI						
8.	ΔP PIPING HWP DISCH TO LOOP				PSI						
	(INCLUDING CONDENSATE FLOW ELEMENT)										
	STATIC HD HWP DISCH TO LOOP				FT						
	DENSITY				LBM/FT IN^2						
9.	ΔP STATIC HWP DISCH TO LOOP				PSI						
	STATIC HD LOOP TO DEMIN INLET HDR				FT						
	DENSITY				LBM/FT IN^2						
10.	ΔP STATIC LOOP TO DEMIN INLET HDR				PSI						
	STATIC HD DEMIN INLET TO OUTLET HDR				FT						
	DENSITY				LBM/FT IN^2						
11.	ΔP STATIC				PSI						
12.	DEMIN ΔP				PSI						

TABLE B-3 (PWR)   CALCULATIONS OF ABSOLUTE PRESSURE – NUCLEAR PLANT – CONDENSATE SYSTEM

	DESCRIPTION				UNITS				0		25	
	FLOW – COND TO NO. 7 DR. IN.				% LBM/HR							
13.	ΔP PIPING DEMIN DISCH TO CBP SUCT				PSI							
	STATIC HD DEMIN DISCH HD TO CBP SUCT				FT							
	DENSITY				LBM/FT IN^2							
14.	ΔP STATIC				PSI							
15.	PRESS AT CBP SUCT (6-7-8-9-10-11-				PSIA							
	12-13-14)											
	DBP TDM				FT							
	DENSITY				LBM/FT IN^2							
16.	CBP PRESS RISE				PSI							
17.	CBP DISCH PRESS (15+16)				PSIA							
18.	ΔP PIPING – CBP to LP7				PSI							
	STATIC HD – CBP to LP7				FT							
	DENSITY				LBM/FT IN^2							
19.	ΔP STATIC				PSI							
20.	ΔP LP7				PSI							
21.	PRESS NO. 7 HTR DR. IN.				PSIA							
	(17-18-19-20)											
22.	ΔP PIPING LP7-LP6				PSI							
	STATIC HD LP7-LP6				FT							
	DENSITY				LBM/FT IN^2							
23.	ΔP STATIC LP7-LP6				PSI							
24.	ΔP LP6				PSI							
25.	ΔP PIPING LP6-LP5				PSI							
	STATIC HD – LP6-LP5				FT							
	DENSITY				LBM/FT IN^2							

TVA 489C (DES-5-48)

TABLE B-3 (PWR)   CALCULATIONS OF ABSOLUTE PRESSURE – NUCLEAR PLANT – CONDENSATE SYSTEM

	DESCRIPTION				UNITS						
	FLOW – COND TO NO. 7 DR. IN.				% LBM/HR				0	25	
26.	ΔP STATIC – LP6–LP5				PSI						
27.	ΔP LP5				PSI						
28.	ΔP PIPING LP5–LP4				PSI						
	STATIC HD – LP5–LP4				FT						
	DENSITY				LBM/FT IN^2						
29.	ΔP STATIC				PSI						
30.	ΔP LP4				PSI						
31.	ΔP PIPING LP4–LP3				PSI						
	STATIC HD LP4–LP3				FT						
	DENSITY				LBM/FT IN^2						
32.	ΔP STATIC				PSI						
33.	ΔP LP3				PSI						
34.	ΔP PIPING LP3 TO NO. 3 DR. IN.				PSI						
	STATIC HD LP3 TO NO. 3 DR. IN.				FT						
	DENSITY										
35.	ΔP STATIC				PSI						
36.	PRESS AT NO. 3 DRAINS IN				PSIA						
	(21–22–23–24–25–26–27–28–29–30–31–										
	32–33–34–35)										
37.	ΔP PIPING – NO. 3 DR. IN. TO MFP SUCT				PSI						
	(INCLUDING SUCTION STRAINER)										
	STATIC HD – NO. 3 DR. IN. TO MFP SUCT				FT						
	DENSITY				LBM/FT IN^3						
38.	ΔP STATIC				PSI						

95

TABLE B-3 (PWR)   CALCULATIONS OF ABSOLUTE PRESSURE - NUCLEAR PLANT - CONDENSATE SYSTEM

	DESCRIPTION			UNITS			0	25
	FLOW - COND TO NO. 7 DR. IN.			% LBM/Hr				
39.	ΔP FLOW ELEMENT IN MFP SUCTION			PSI				
40.	PRESS AT MFP SUCT (36-37-38-39)			PSIA				
41.	VAPOR PRESS - MFP SUCT			PSI				
42.	MFP NPSH			PSIA				
43.	DENSITY			LBM/FT IN^2				
44.	MFP NPSH			FT				
45.	TOTAL COND. TDH REQ'D			PSI				
	(40+2+3+7+8+9+10+11+12+13+14+							
	18+19+20+22+23+24+25+26+27+							
	28+29+30+31+32+33+34+35)							
	DENSITY			LBM/FT^3				
46.	TOTAL COND TDH REQ'D			FT				

TABLE B-4 (BWR)   CALCULATIONS OF ABSOLUTE PRESSURE – NUCLEAR PLANT – CONDENSATE SYSTEM

	DESCRIPTION				UNITS					
	FLOW – COND TO NO. 3 DR. IN.				% LBM/HR			0		25
1.	CONDENSER PRESSURE – HP ZONE				PSIA					
	STATIC HD – NWL TO HWP SUCT FLG				FT					
	DENSITY				LBM/FT IN^2					
2.	ΔP STATIC				PSI					
3.	ΔP PIPING COND TO HWP SUCT				PSI					
	(INCLUDING SUCTION STRAINER)									
4.	HWP SUCTION PRESS (1-2-3)				PSIA					
	HWP TDH				FT					
	DENSITY				LBM/FT IN^2					
5.	HWP PRESS RISE				PSI					
6.	HWP DISCH PRESS (4+5)				PSIA					
	HEAD LOSS SJAE, OGC, GSC				FT					
	DENSITY									
7.	ΔP SJAE, OGC, GSC				PSI					
8.	ΔP PIPING HWP DISCH TO SJAE, OGC,				PSI					
	GSC LOOP (INCLUDING COND FLOW ELEMENT)									
	STATIC HD HWP DISCH TO LOOP				FT					
	DENSITY				LBM/FT IN^2					
9.	ΔP STATIC HWP DISCH TO LOOP				PSI					
	STATIC HD LOOP TO DEMIN INLET HDR				FT					
	DENSITY				LBM/FT IN^2					
10.	ΔP STATIC LOOP TO DEMIN INLET HDR				PSI					
	STATIC HD DEMIN INLET TO OUTLET HDR				FT					
	DENSITY				LBM/FT IN^2					
11.	ΔP STATIC				PSI					
12.	DEMIN ΔP				PSI					

TABLE B-4 (BWR)   CALCULATIONS OF ABSOLUTE PRESSURE – NUCLEAR PLANT – CONDENSATE SYSTEM

	DESCRIPTION					UNITS						0	25
	FLOW – COND TO NO. 3 DR. IN.					% LBM/HR							
13.	ΔP PIPING DEMIN DISCH TO CBP SUCT					PSI							
	STATIC HD DEMIN DISCH HDR TO CBP SUCT					FT							
	DENSITY					LBM/FT IN^2							
14.	ΔP STATIC					PSI							
15.	PRESS AT CBP SUCT (6-7-8-9-10-11-12-13-14)					PSIA							
	CBP TDH					FT							
	DENSITY					LBM/FT IN^2							
16.	CBP PRESS RISE					PSI							
17.	CBP DISCH PRESS (15+16)					PSIA							
18.	ΔP PIPING – CBP TO DC7					PSI							
	STATIC HD – CBP TO DC7					FT							
	DENSITY					LBM/FT IN^2							
19.	ΔP STATIC					PSI							
20.	ΔP DC7					PSI							
21.	ΔP PIPING DC7 TO LP7					PSI							
	STATIC HD – DC7 TO LP7					FT							
	DENSITY					LBM/FT IN^2							
22.	ΔP STATIC					PSI							
23.	ΔP LP7					PSI							
24.	ΔP PIPING LP7 – LP6					PSI							
	STATIC HD – LP7-LP6					FT							
	DENSITY					LBM/FT IN^2							
25.	ΔP STATIC					PSI							
26	ΔLP6					PSI							

TABLE B-4 (BWR)　　　CALCULATIONS OF ABSOLUTE PRESSURE - NUCLEAR PLANT - CONDENSATE SYSTEM

	DESCRIPTION				UNITS				0	25	
	FLOW - COND TO NO. 3 DR. IN.				% LBM/HR						
27.	ΔP PIPING LP6-LP5				PSI						
	STATIC HD - LP6-LP5				FT						
	DENSITY				LBM/FT IN^2						
28.	ΔP STATIC - LP6-LP5				PSI						
29.	ΔP LP5				PSI						
30.	ΔP PIPING LP5-LP4				PSI						
	STATIC HD - LP5-LP4				FT						
	DENSITY				LBM/FT IN^2						
31.	ΔP STATIC				PSI						
32.	ΔP LP4				PSI						
33.	ΔP PIPING LP4-LP3				PSI						
	STATIC HD LP4-LP3				FT						
	DENSITY				LBM/FT IN^2						
34.	ΔP STATIC				PSI						
35.	ΔP LP3				PSI						
36.	ΔP PIPING LP3 TO NO. 3 DR. IN.				PSI						
	STATIC HD LP3 TO NO. 3 DR. IN.				FT						
	DENSITY				LBM/FT IN^3						
37.	ΔP STATIC				PSI						
38.	PRESS AT NO. 3 DRAINS IN				PSIA						
	(17-18-19-20-21-22-23-24-25-										
	26-27-28-29-30-31-32-33-34-35-										
	36-37)										

TABLE B-4 (BWR)   CALCULATIONS OF ABSOLUTE PRESSURE – NUCLEAR PLANT – CONDENSATE

	DESCRIPTION				UNITS					0	25
	FLOW – COND TO NO. 3 DR. IN.				% LBM/HR						
39.	ΔP PIPING – NO. 3 DR. IN. TO MFP SUCT (INCLUDING SUCTION STRAINER)				PSI						
	STATIC HD – NO. 3 DR. IN. TO MFP SUCT				FT						
	DENSITY				LBM/FT IN^2						
40.	ΔP STATIC				PSI						
41.	ΔP FLOW ELEMENT IN MFP SUCTION				PSI						
42.	PRESS AT MFP SUCT (35-36+37-38)				PSIA						
43.	TOTAL COND TDH REQ'D (42+2+3+7+ 8+9+10+11+12+13+14+18+19+20+22+ 23+24+25+26+27+28+29+30+31+32+33+ 34+35+37+38+39)				PSI						
44.	DENSITY				LBM/FT IN^2						
45.	TOTAL COND TDH REQ'D				FT						
46.	VAPOR PRESS – MFP SUCT				PSIA						
47.	MFP NPSH (42-46)				PSI						
	DENSITY				LBM/FT IN^2						
	MFP NPSH				FT						

100

Chapter 4
Design Guide DG-M2.4.1: Extraction Steam System

1.0 GENERAL

   1.1 SCOPE

This design guide is an aid for the design of the extraction steam system of a 3800 megawatt-thermal (MWT) nuclear plant with a light water cooled reactor.

Use of this design guide permits preparation of a draft design criteria for a specific project. An engineer with a thorough understanding of the system should prepare the design criteria and perform the design calculations. Appendix A includes a schematic and the necessary calculations for the basic system design. Neither the appendix nor the schematic should be a part of the design criteria. Safety-related functions are not treated in this document.

   1.2 APPLICABLE CODES AND STANDARDS

All piping in the system shall be designed in accordance with ANSI B31.1, "American National Standard Code for Pressure Piping--Power Piping." All valves shall be designed in accordance with ANSI B16.34, "American National Standard--Steel Valves." Any flanged fittings shall be designed in accordance with ANSI B16.5, "American National Standard--Steel Pipe Flanges and Flanged Fittings

2.0 FUNCTIONAL DESCRIPTION

The extraction steam system shall supply the steam required for feedwater heating and for the moisture separator reheater (MSR) first-stage reheater. The extraction steam system boundaries extend from the connection on the main turbine or turbine cold reheat piping to the feedwater heaters and MSR first-stage reheater. The unit will have seven stages of feedwater heating arranged in three parallel streams The seven stages of feedwater heating will be numbered sequentially in the order of decreasing heater shell pressure. The three streams of feedwater heaters are designated by letters, the highest pressure heaters being 1A, 1B, and 1C.

The extraction steam system (in a BWR system only) shall also supply the required heating steam for the steam seal evaporator.

The functional requirements of the system are met by providing the components described in the paragraphs which follow.

   Extractions No. 1, No. 2, and No. 3 will each be taken from one or more nozzles on the turbine (or cold reheat piping) and combined into a single extraction line supplying all the feedwater heaters of each stream for that extraction. Each line branches into three extraction lines, one for each feedwater heater. Steam for the

MSR first-stage reheater will also be taken from the high-pressure turbine. Motor-operated isolation valves shall be provided in the extraction lines for each feedwater heater and each MSR first-stage reheater. (When No. 3 extraction is taken from the MSR shell drain, no isolation valves will be provided.) Positive closing check valves shall be provided in the extraction lines to each No. 1 and No. 2 feedwater heater and each MSR first-stage reheater. Steam for the building heating, ventilating, and air-conditioning (HVAC) system and radwaste evaporator shall be supplied from cold reheat.

For a Brown Boveri Turbogenerator with a PWR, an Overload Valve (OLV) may supply main steam to extraction No. 1 for "stretch" unit load conditions. For a Brown Boveri Turbogenerator with a BWR, an OLV may supply main steam to the first stage reheater for "stretch" load conditions.

Extraction No. 4 will be taken from the turbine by a separate extraction line for each heater, and will not be interconnected with the heaters of the other two streams. A motor-operated isolation valve and positive closing check valve shall be provided in the extraction line to each heater.

Feedwater heaters No. 5, No. 6, and No. 7 will be mounted in the condenser neck. Extraction lines to feedwater heaters of separate streams will not be connected, and will not contain any isolation valves. Antiflash baffles shall be provided in these feedwater heaters or Mission Duo-Check valves shall be provided in the extraction lines for the prevention of turbine overspeed.

If any heater is connected to a source of steam which is external to the turbine cycle (such as pegging steam from the steam generator, or reactor for a BWR, or auxiliary boiler system), the extraction line to that heater shall have two positive closing check valves in series. The second valve may be either in the individual heater extraction lines or in the common extraction line (whichever arrangement is more economical).

3.0 SYSTEM DESIGN REQUIREMENTS

The extraction steam system shall operate only when the main turbine is in operation or starting up. The extraction steam system shall be capable of operation with one section of one stream of feedwater heating out of service without reducing turbine load. The layout of extraction piping within the condenser neck shall be coordinated with the manufacturers of the condenser and turbine.

4.0 DESIGN INFORMATION

4.1 COMPONENTS

4.1.1 Valves

4.1.1.1 <u>PWR only</u>--Motor-operated isolation valves shall be provided in each extraction line to feedwater heaters No. 1, No. 2, No. 3, and No. 4 and to the MSR first-stage reheater. These valves shall be capable of manual actuation from either the control room or local panels. Once these valves begin to close, they shall not be permitted to stop or reopen until the valve has closed completely. Limit switches shall be provided for control room indication that these valves are in the open or closed position.

4.1.1.2 <u>BWR only</u>--Motor-operated isolation valves shall be provided in each extraction line to feedwater heaters No. 1 and No. 2. These valves shall be capable of manual operation from either the control room or local panels to control heater pressure (thus feedwater temperature). This allows maximum fuel burnup during the end of the fuel cycle. Limit switches shall be provided for control room indication that these valves are in the open or closed position.

Motor-operated isolation valves shall be provided in each extraction line to feedwater heaters No. 3 and No. 4 and to the MSR first-stage reheater. These valves shall be capable of manual actuation from either the control room or local panels. Once these valves begin to close, they shall not be permitted to stop or reopen until the valve has closed completely. Limit switches shall be provided for control room indication that these valves are in the open or closed position.

4.1.2 Piping

4.1.2.1 Each extraction line shall have the same design pressure as the corresponding feedwater heater.

4.1.2.2 Piping design shall include considerations of nozzle loading for the main turbine and feedwater heaters. All extraction piping shall be arranged, where possible, to be continuously sloped from the turbine connection to the point of discharge to the feedwater heaters. Where this arrangement is not possible, or where condensate can accumulate ahead of a closed valve, low-point drains shall be provided to ensure a dry extraction piping system.

4.1.2.3 Low-point drains shall be either orificed or trapped and should operate continuously. Drain pots (or enlarged pipe sections) shall be provided at the drain connection to the extraction pipe to permit installation of a level switch for indication of drain malfunction. Low-point drains shall also be equipped with a bypass valve for use during startup and low-load operation.

The startup drains shall be sized to pass the required condensate flow for warmup of the main turbine, with static head as the only driving potential.

4.1.2.4 Provisions shall be made ahead of the feedwater heater isolation valves for removal of water if a low point in the piping is formed when a valve is in the closed position.

4.1.2.5 All extraction lines except No. 3 shall be designed for a maximum fluid velocity of 150 feet per second at 100 percent load. Due to the excessive moisture in the steam, the No. 3 extraction line shall be designed for a maximum fluid velocity of 100 feet per second at 100 percent load. The extraction line shall be made of an erosion-resistant steel alloy (ASTM A-335, Grade P22, or other copper-nickel-chromium alloy with total alloy composition of approximately 1 percent by weight).

4.1.2.6 All extraction piping containing wet steam with moisture in excess of 5 percent shall have an erosion allowance above the minimum wall thickness required for design pressure.

4.1.2.7 For BWR systems and Babcock and Wilcox NSSS only, the first-stage reheater tube bundle, if equipped with carbon steel tubes, shall have provisions for warming from the auxiliary steam system to prevent corrosion of the carbon steel tubes during shutdown.

4.1.2.8 Provisions shall be made in the heater drains and vents system for removal of moisture from the low pressure turbines while a heater stream is isolated (see DG-M2.5 "Heater Drains and Vents" for details).

4.1.3 Instrumentation

4.1.3.1 In addition to the instrumentation required to perform the logical functions described in this guide, instrumentation shall be provided to: (a) measure system temperatures and pressures required to satisfy the input needs of the balance of plant computer system, (b) provide test information required under ASME PTC-6, and (c) give additional information about operation of the system equipment.

4.1.3.2   If the output of the sensing elements is not required for plant computer or logic functions, output display on local panels shall be sufficient.

Control room indication is required for:

a.   Pressure in each feedwater heater shell.

b.   Extraction steam temperature in piping with superheated steam.

All pressure monitoring connections shall be equipped to facilitate on-line instrument calibration without disturbing station instrumentation.

4.1.3.3   Injection and sampling connections for a radioactive tracer shall be located in the extraction lines requiring the measurement of wet steam enthalpy. The sampling connection shall be located to prevent entrainment of steam.

4.1.4   Insulation

4.1.4.1   All piping and equipment outside the condenser whose temperature may exceed 150° F shall be insulated so that surface temperatures do not exceed 125° F with ambient air at 80° F.

4.1.4.2   Extraction piping in the condenser neck shall be wrapped with a sheathing of 0.06-inch, 12 percent chromium steel or 18-8 stainless steel to form a concentric annular air space of at least 1 inch radial thickness. Provisions shall be made for adequate drainage while minimizing air or steam circulation in the annular space.

4.1.5   Heater Design Pressure

Shell design pressure for feedwater heaters shall be equal to 1.15 times the turbine stage pressure at 100 percent load, rounded to the next highest 10 psi. The minimum design pressure of any heater is 50 psi.

4.2   LOGIC FOR OPERATION

4.2.1   All extraction line positive-closing check valves shall be initiated toward the closed position on main turbine trip.

4.2.2   Each extraction steam isolation valve and positive-closing check valve shall be closed on emergency high water level in its respective feedwater heater isolation group.

The positive-closing check valves shall be capable of remote actuation from the control room and local control panel.

4.2.3 Extraction steam isolation valves in the line to the MSR first-stage reheater shall close automatically on indication of an emergency (high-high) level in the respective first-stage reheat drain tanks.

4.2.4 All low-point drain bypasses shall be controlled from a single switch in the control room. Capability for controlling individual bypass drain valves shall also be provided locally. If the bypass valve is not readily visible, local indication shall be provided to indicate that the valve is in the closed position.

4.3 SYSTEM INTERFACES

The following EN DES design guides will assist in defining system interfaces and provide additional insight into the design of an extraction steam system for a 3800 MWT nuclear plant with light water reactors:

a. DG-M2.1, "Main and Reheat Steam System"

b. DG-M2.2, "Main Feedwater System"

c. DG-M2.3, "Condensate System"

d. DG-M2.5, "Heater Drains and Vents"

e. DG-M2.7, "Condenser and Auxiliaries"

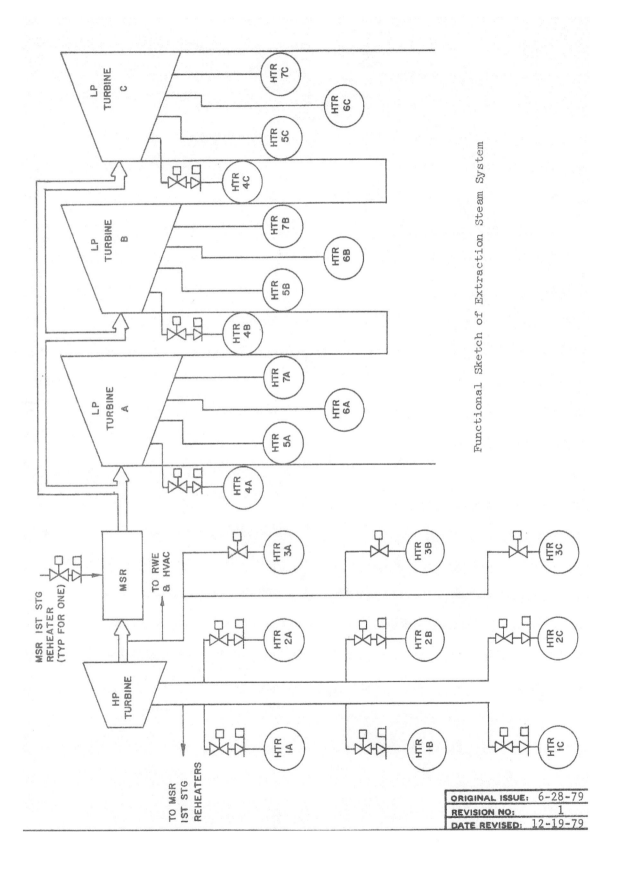

Functional Sketch of Extraction Steam System

ORIGINAL ISSUE:	6-28-79
REVISION NO:	1
DATE REVISED:	12-19-79

APPENDIX A

DESIGN CALCULATIONS

FOR

EXTRACTION STEAM SYSTEM

1.0 PURPOSE

This design guide presents the necessary calculations for the basic
design of the extraction steam system. It sets forth a format for
these calculations to insure that the system design is consistent
among TVA plants and the calculations are a convenient source of
reference.

2.0 NECESSARY INFORMATION

The following information is necessary to perform the extraction
steam system calculations:

a. From the turbine generator contractor--Turbine generator heat
   balance at 100 percent of unit load. This heat balance will be
   used primarily to obtain extraction steam conditions and flows.

b. From the task force plant layout and piping group--Piping isometric
   sketches showing lengths between changes in pipe diameter or
   branches in piping and approximate location of valves and fittings.
   The isometric should show the locations of the extraction steam
   nozzles on the turbine generator and feedwater heaters. The plant
   layout and equipment group should refer to the design criteria
   diagram for extraction steam piping requirements. The extraction
   steam system sketch is included in this design guide as an aid in
   developing the design criteria diagram.

3.0 EXTRACTION STEAM PIPE SIZING

3.1 Calculations should be performed to obtain the pressure drops
    in the extraction steam lines at 100 percent NSSS rating. The
    calculations should consider pipe friction loss and head
    losses in fittings and valves. Mechanical Design Guide
    DG-M2.11, Pressure Drop Calculations for Steam and Condensate
    Piping and Fittings, should be used for making these calculations.

3.2 Pipe sizes should be chosen to limit extraction line pressure
    drops to five percent of the pressure at the extraction nozzle
    connection on the turbine-generator. Steam velocities should be
    limited to 150 feet per second in all extraction lines, except
    extraction No. 3. Steam velocity shall be limited to 100 feet

108

per second in extraction No. 3 to minimize erosion due to the excessive moisture. Internal design pressure of the piping shall be equal to the shell side design pressure of each respective feedwater heater.[1]

3.3 The piping isometric sketches from the plant layout and piping group should be used to obtain piping lengths, number of fittings, and other related information.

---

1. Since a special alloy is used, table 2 of design guide DG-M2.11 is not applicable for the No. 3 extraction line.

Chapter 5
Design Guide DG-M2.5.1: Heater Drains and Vents System

1.0 GENERAL

This design guide serves as an aid for designing non-safety-related heater drains and vents systems for 3800 MW nuclear plants having light-water reactors.

Assistance is given for the preparation of project design criteria drafts. Sketches of typical system configurations for both boiling water reactor (BWR) and pressurized water reactor (PWR) plants are displayed in Appendix A. These may be used to develop Design Criteria Diagrams. Methods for preparing design calculations and establishing performance criteria necessary for the procurement of heater drain pumps are outlined in Appendix B.

A discussion of safety-related functions is not within the scope of this document.

1.1 ABBREVIATIONS AND SYMBOLS

APC	- Atmospheric pressure condensate drain tank
BOP	- Balance of plant
BWR	- Boiling water reactor
CBP	- Condensate booster pump
C_v	- Valve flow coefficient
cv	- Control valve
$C.V.\Delta P_{all}$	- Control valve allowable pressure drop, lb/in^2
G	- Pump flow, gal/min
ΔP	- Pressure drop, lb/in^2
ΔP_{all}	- Allowable pressure drop, lb/in^2
H_X	- Pressure drop in lb/in^2 at load X
H_{100}	- Pressure drop in lb/in^2 at 100% load
HD	- Heater drain
HDP	- Heater drain pump

111

HDT	-	Heater drain tank
HVAC	-	Heating, ventilation, and air-conditioning
HWP	-	Hotwell pump
K_m	-	Valve recovery coefficient
lb/in^2	-	Pounds per square inch
lb/in^2a	-	Pounds per square inch, absolute
lb/in^2g	-	Pounds per square inch, gauge
LCV	-	Level control valve
min	-	Minimum
max	-	Maximum
mfp	-	Main feed pump
mfpt	-	Main feed pump turbine
ms/rh	-	Moisture separator reheater
N	-	Number of drain pumps operating
npsh	-	Net positive suction head
NSSS	-	Nuclear steam supply system
OGC	-	Off-gas condenser
OGS	-	Off-gas system
P_1	-	Control valve inlet pressure, lb/in^2a
P_v	-	Vapor (saturation) pressure of the fluid, lb/in^2a
PWR	-	Pressurized water reactor
Q_X	-	Flow at load point X in lb/hr
Q_{100}	-	Flow at 100% load in lb/hr
r_c	-	Critical pressure ratio
sjae	-	Steam jet air ejector

SV   - Specific volume

TEFC   - Totally enclosed fan cooled

TEWAC   - Totally enclosed water and air cooled

W   - Total heater drain flow, lb/hr

1.2 APPLICABLE CODES AND STANDARDS

All pressurized components within a system will be designed and
constructed to meet the requirements outlined in ANSI B31.1,
"American National Standard Code for Pressure Piping--Power
Piping," and the ASME Boiler and Pressure Vessel Code,
Section VIII, Division 1. Feedwater heaters will be
designed and constructed to comply with the Heat Exchange
Institute's "Standard for Closed Feedwater Heaters," and the
system pump design with the Standards of the Hydraulic Institute.

2.0 FUNCTIONAL DESCRIPTION

A heater drains and vents system performs the following functions
during normal plant operating conditions:

a. Regulates heater shell, heater drain tank, moisture separator drain
    pot, and reheater drain tank condensate levels.

b. Returns all condensate coming from the feedwater heaters, moisture-
    separator reheaters (ms/rh's), and heat exchangers to the condensate
    system.

c. Vents the feedwater heaters, ms/rh's, and heat exchangers to
    maintain the satisfactory removal of noncondensable gases.

d. Provides equipment maintenance drains by means of a totally
    enclosed piping system to the main condenser hotwell (BWR only).

A heater drains and vents system extends from the drain and vent
connections on the feedwater heaters, ms/rh's, or other heat exchangers
to the main condenser hotwell or connection in the condensate system
piping. A heater drains and vents system contains the instrumentation
and controls associated with its components and piping.

Heater drains are designated like feedwater heaters. The seven stages
of feedwater heating are numbered sequentially in order of decreasing
heater shell pressure. The three streams of feedwater heaters are
designated sequentially by letters (i.e., the highest pressure feedwater
heaters are 1A, 1B, and 1C.)

Flow diagrams depicting both "high-pressure" and "low-pressure" (BWR and PWR) heater drains and vents systems are included in Appendix A.

Each No. 1 heater has a normal operating drain routed to a No. 2 heater in the same stream (i.e., heater 1A drains to 2A, 1B to 2B, and 1C to 2C). Each No. 2 heater has a normal operating drain routed to a No. 3 heater drain tank. In addition, each No. 1 heater has a bypass drain routed to each No. 2 heater, and each No. 2 heater has a bypass drain directed to a condenser.

Each No. 3 feedwater heater is a dry shell type draining to a common drain tank and having no drain cooler. The No. 3 heater drain tank is shared by all three strings of feedwater heaters. It receives drains from the No. 3 heaters, the No. 2 heaters, the moisture-separators, and the main feed pump intermediate leakoff. Drain tank condensate level is controlled inside the No. 3 heater drain tank. The normal operating drain from the No. 3 heater drain tank is pumped forward into the condensate system downstream from the No. 3 feedwater heater. In addition, the No. 3 heater drain tank has a bypass drain directed to the condenser. Steam that flashes from water inside the No. 3 heater drain tank is vented to each No. 3 heater shell.

Each No. 4 heater has a normal operating drain routed to the No. 5 heater having the same stream and a bypass drain going to the main condenser. Each No. 5 heater has a normal operating drain and a bypass drain routed to the No. 6 heater of the same stream. (Heater No. 4A drains to 5A, 4B to 5B . . . 5A to 6A, etc.)

PWR Only--Each No. 6 heater has a normal operating drain routed to a No. 7 heater drain tank and a bypass drain going to the main condenser. Each No. 7 heater is the dry shell type that has no drain cooler and drains to a common drain tank. Drain tank condensate level is controlled inside the No. 7 heater drain tank. The No. 7 heater drain tank is shared by all three strings of feedwater heaters. The normal operating drain coming from the No. 7 heater drain tank is pumped into the condensate system downstream from the No. 7 feedwater heaters. In addition, the No. 7 heater drain tank has a bypass drain to the condenser. Steam that flashes from water inside the No. 7 heater drain tank is vented to each No. 7 heater shell.

All heater-shell-side relief valves and channel-side sentinel valve discharges are piped to equipment drain funnels that are connected to the atmospheric condensate drain tank. Maintenance drains for both channel and shell sides of the heaters must be installed.

BWR Only--Each No. 6 heater has a normal operating drain and a bypass
    drain routed to a No. 7 heater located in the same stream.

    Each No. 7 heater has a separate external level control reservoir and
    drain cooler with the drain tank condensate level controlled inside
    the level control reservoir. Each No. 7 heater has a normal operating
    drain and a bypass drain routed through the drain cooler to the
    condenser.

    All relief valve discharges are piped to the main condenser. Main-
    tenance drains for both channel and shell sides of the heaters must
    be installed. The steam jet air ejector (sjae) condensers, off-gas
    condenser (OGC), and off-gas preheaters drain directly to the main
    condenser hotwell.

All feedwater heaters have continuous vents running to the main condenser to be used during operation of the plant.

Maintenance drains (i.e., for heaters No. 4, No. 5, and No. 6) are
sized to permit moisture removal from the turbine internal moisture
separator, during low load operation or when the heaters are isolated.
A bypass to the condenser must be installed around the manual isolation
valves along with a motor-operated valve and an orifice.

The ms/rh's have moisture separator drain tanks, first-stage reheater
drain tanks, and second-stage reheater drain tanks. The moisture
separator drains are transported to the No. 3 heater drain tank. The
first-stage reheater drains are transported to either heater No. 1 or
heater No. 2, and the second-stage reheater drains go to heater No. 1.
The reheaters are vented to the turbine extraction lines that are routed
to the No. 1 and/or No. 2 heaters depending on extraction pressures.
When the moisture separator drain is used as a primary source for steam
extraction the unit may be designed without a moisture separator drain
tank.

Level control valves control the drains during normal operation. Bypass
level control valves that are routed to the condenser permit continued
operation of the system in the event of component malfunction, and they
facilitate system startup.

The drain tank condensate level in the first-stage reheater drain tanks is
controlled by level control valves in the drain going to the No. 1 or
No. 2 feedwater heaters. The drain tank condensate level in the second-
stage reheater drain tanks is controlled by the level control valves
located in the drain to the No. 1 feedwater heaters.

Drains, coming from the steam seal condenser and main feed pump (mfp)
outboard leakoff, flow to an atmospheric pressure condensate drain tank
that drains to the main condenser. Drains from the mfp intermediate
leakoff go to the No. 3 HDT.

3.0 SYSTEM DESIGN REQUIREMENTS

3.1 STARTUP AND NORMAL OPERATION TO MAXIMUM ANTICIPATED POWER LEVEL

During the startup operation, all heater drains go to the condenser to satisfy feedwater quality. When the unit load exceeds 40 percent, and the feedwater quality requirements are satisfied, the heater drain pumps begin delivering drain flow to the condensate system.

The vent system evacuates the heaters, drain tanks, and reheaters during startup, before admitting steam to the turbine, and removes noncondensibles from the heaters, drain tanks, and reheaters during normal power operation.

3.2 TRANSIENT CONDITIONS

The system is designed to operate satisfactorily while interfacing with the feedwater and condensate systems, and maintain proper level control and venting during the transient conditions that follow:

a. Main turbine trip.

b. Loss of a section of one stream of feedwater heaters without reducing turbine load.

c. Loss of one heater drain pump without reducing turbine load.

d. Loss of one main feed pump without reducing turbine load.

e. Loss of all HDP's resulting in the drain flow being bypassed to the condenser (accompanied by a turbine load reduction).

f. Failure of one HDP level control valve.

g. MSR operation under abnormal or test conditions.

4.0 SYSTEM DESIGN INFORMATION

4.1 PERFORMANCE CRITERIA

4.1.1 Feedwater Heaters

The Nos. 1, 2, 4, 5, and 6 feedwater heaters are designed to have a drain cooler approach temperature of 10°F. The No. 7 heater drain cooler for a BWR will also have a drain cooler approach temperature of 10°F.

116

The heater drain nozzles and flow passages are sized to accommodate 150 percent of the drain flow that occurs during normal operation at full load.

4.1.2   Heater Drain Tanks

a.   No. 3 Heater Drain Tank Size

The No. 3 heater drain tank is sized for a minimum storage capacity of 1 minute (based on normal tank water level) when the total flow reaches 105 percent of the NSSS rating. A vent routed back to the No. 3 heaters is necessary to prevent backpressure in the drain tank caused by the flashing of higher pressure drains that could prevent drainage of the No. 3 heaters. The vent is sized to accommodate 100 percent of the flashing steam at 105 percent of the NSSS rating with a pressure drop which is at least 3 lb/in^2 less than the static pressure between the bottom of the No. 3 heaters and the hi-hi water level in the No. 3 HDT.

b.   No. 7 Heater Drain Tank Size

The No. 7 heater drain tank for a PWR is sized for a minimum storage capacity of 1 minute, based on normal tank water level that occurs when the total flow from the No. 7 heaters and the No. 6 heaters approaches 105 percent of the NSSS rating. A vent routed back to the No. 7 heaters is needed to prevent backpressure in the drain tank (caused by the flashing of the No. 6 heater drains) that could prevent drainage of the No. 7 heaters. The vent is sized to accommodate 100 percent of the flashing steam at 105 percent of NSSS rating with a pressure drop which is at least 3 lb/in^2 less than the static pressure between the bottom of the No. 7 heaters and the hi-hi water level in the No. 7 HDT.

The normal operating level is based on design calculations that determine tank size, npsh and flow capacitance. Normal operating level is situated above the centerline of the tanks to minimize tank size and obtain the maximum available npsh for the drain pumps.

Sufficient static head must be available from the drain tank normal operating level to the condenser connection to allow sufficient drainage during unit startup when no pressure differential exists between heater drain tank and the condenser.

117

4.1.3   Heater Drain Pumps

The three No. 3 heater drain pumps and the two No. 7 heater
drain pumps (PWR) are rated and designed to operate at peak
efficiency for the flow required at 100 percent turbine load.
The required head at rated conditions is selected so that the
pumps can meet all of the following conditions, assuming the
pump has the lowest practical head rise to shutoff.

a.   Three No. 3 heater drain pumps and the two No. 7 heater drain
     pumps must supply sufficient flow and head to pump all drains
     forward at 100 percent load. This occurs when three main feed
     pumps, three hotwell pumps, three condensate booster pumps and
     all feedwater heaters are in service; there is a maximum
     pressure drop across the demineralizers.

b.   The three No. 3 heater drain pumps and the two No. 7 heater
     drain pumps (PWR) supply the flow and head required at
     105 percent turbine load. This occurs when three main
     feed pumps, three hotwell pumps and all feedwater heaters
     are in service; there is a minimum pressure drop across
     the demineralizers.

c.   Two No. 3 heater drain pumps and one No. 7 heater drain pump
     (PWR) supply the flow and head required to begin pumping
     heater drains forward at a turbine load of from between 40
     to 50 percent of the turbine rating. This occurs when three
     main feed pumps, three hotwell pumps, three condensate booster
     pumps, and all feedwater heaters are in service; there is a
     minimum pressure drop across the demineralizers.

4.1.4   Level Control Valves

Heater Drain Control Valves--With the exception of level control
valves associated with the heater drain pumps, all other level
control valves are sized to be fully open for 150 percent of
the drain flow existing at full load during normal operation.
Since heater drain flows having one stream out of service do
not reach 150 percent of the flow existing during normal operation, no other allowance for dynamic response is required.
Both valves used during normal operation and the bypass level
control valve are capable of controlling a stable level at
10 percent of unit rating. For all bypass level control valves,
sufficient static head must be available between the heater
level and the downstream heater or condenser connection to
assure drainage during startup. Sufficient static head between
the heater level and the bypass level control valve prevents
flashing upstream of the valve seat.

118

No. 3 Heater Drain Control Valves--The two No. 3 heater drain pump control valves are sized to ensure stable operation with the heater drains pumped forward from approximately 40 percent to 105 percent of turbine rated load. One valve is sized to supply sufficient backpressure to prevent pump runout if one or more heater drain pumps are out of service.

The No. 3 drain tank bypass LCV to the main condenser is sized to accommodate the total drain flow at 50 percent turbine load. The bypass LCV must also control a stable level down to 10 percent turbine rating. Sufficient head must be available between the water level in the tank and its corresponding bypass level control valve to prevent flashing upstream of the valve seat.

No. 7 Heater Drain Control Valves

PWR Only--The No. 7 heater drain pump control valve is sized
    to ensure stable operation with the heater drains pumped
    forward from approximately 25 percent to 105 percent of
    turbine rated load. This valve is sized to accommodate all
    of the drain flow during 105 percent turbine load with two
    pumps operating. This valve also accommodates all of the
    drain flow during 50 percent turbine load when one pump is
    operating.

BWR Only--The No. 7 heater drain control valve is sized to
    accommodate all of the drain flow for 100 percent turbine
    load at approximately 85 percent of the total travel. The
    valve must be capable of stable level control down to
    10 percent of turbine rating.

Atmospheric Drain Tank Control Valves--A level control valve is installed in the drain line routed from the atmospheric pressure condensate drain tank (APC) to the condenser. In addition, a bypass level control valve is installed in parallel. Each of these valves is sized to accommodate the maximum flow expected from the atmospheric pressure condensate drain tank at approximately 85 percent of total travel. These valves must be capable of stable level control down to 10 percent of the turbine rating.

MSR Drain Control Valves--The criteria for sizing the moisture separator drain control valves and the reheater drain control valves are determined by the turbine generator manufacturer.

4.2 COMPONENT CRITERIA

4.2.1 Feedwater Heaters

This design guide establishes the requirements for the shell side of feedwater heaters. Channel side requirements are discussed in section 4.5.9 of design guide DG-M2.2.1, "Main Feedwater System."

PWR Only--Each feedwater heater is a closed, horizontal, U-tube type with carbon steel shells. Heaters 5, 6, and 7 are mounted in the condenser neck.

BWR Only--The high-pressure feedwater heaters for a BWR are of the closed, vertical, channel down, U-tube type. The low-pressure feedwater heaters are of the closed, horizontal, U-tube type. Heaters 5, 6, and 7 are mounted in the condenser neck. The shells of all the heaters are constructed of carbon steel.

Shell side relief valves are included on each No. 1, No. 2, No. 3, and No. 4 heaters.

Maintenance drains and vents are included on each heater shell and channel to facilitate condensate drainage when the heater is out of service.

Each heater has continuously operating vents, including the necessary orifices. Startup vent connections are not needed, since vacuum conditions are present for long periods of time before turbine startup.

The feedwater heater shells are designed for vacuum and design pressure conditions. The design pressure is 115 percent of the extraction pressure during 100 percent turbine load. The design pressure for the No. 3 heater is the same as the MSR shell design pressure.

The design temperature for the heaters is the saturation temperature at the design pressure; or the steam temperature during 105 percent turbine load, whichever is greater.

The feedwater heaters include nozzles for extraction, a drain, a level control instrument header, and vent lines when required.

The Nos. 1, 2, 4, 5, and 6 heaters must have separate connections to the instrument taps for the normal operating drain level controls and the bypass level controls unless the level instrument connections are at least 3 inches in diameter.

120

The drain cooler section of the heaters, when required, is designed with the inlet opening sized to prevent the flashing of saturated water in the drain cooler inlet when the heater is at minimum water level.

4.2.2   Heater Drain Tanks

The tanks are constructed of carbon steel and designed in accordance with ASME B&PV Code Section VIII, Division 1; and have the same design pressure and temperature as their respective heaters.

The heater drain tanks have nozzles for: each incoming drain; vent lines to their respective heaters; a drain pump suction line; each drain pump minimum recirculation line; each bypass drain line; and all instrumentation. Level instrument connections must be located at least 6 inches above the tank bottom.

The drain tanks have internal baffle plates for all flashing inlet drains. An internal wave dampening plate must be installed in the liquid portion of the tank to separate the tank level control area from the drain inlet area.

The drain pump suction connection on the drain tank must protrude upward 6 inches from the inside bottom of the tank; and a baffle plate should be installed above the connection to prevent objects from entering the pump suction area. The bypass drain connection must be flush with the bottom of the drain tank.

The tanks must have separate connections to instrument taps for the normal operating drain level control and the bypass level control, unless the level instrument connections are at least 3 inches in diameter.

4.2.3   Heater Drain Pumps

The heater drain pumps are centrifugal, horizontal or vertical type, depending on the npsh available and the turbine building arrangement.

Heater drain pumps are designed to meet the Hydraulic Institute Standards. The pump discharge side design pressure is the maximum discharge pressure when the pump is operating at minimum recirculation flow, and when the heater drain tank is operating with the maximum operating water level and pressure. The pump suction side design pressure is the design pressure of the HDT plus the static pressure from the maximum operating water level inside the HDT to the pump suction. The design temperature for the heater drain pumps is the same as for their respective drain tanks.

The heater drain pumps are equipped with mechanical seals having condensate injection. Each heater drain pump has drain, vent, and cooling water connections.

Each heater drain pump is driven by a 6600-volt, 3-phase, 60-Hertz motor. Each motor is equipped with six thermocouples (two per phase; one is a backup) for the windings and one thermocouple for each bearing. High-temperature control room alarms should be installed for each motor thermocouple. Motors may be totally enclosed, fan cooled (TEFC) or totally enclosed water and air cooled (TEWAC), depending on an economic evaluation.

Each pump and motor should be equipped with a self-contained lubricating oil system providing forced-feed lubrication to all pump and motor bearings. The system should include:

a. A shaft-driven oil pump.

b. A separate 230/460-volt, 3-phase, 60-Hertz, motor-driven oil pump.

c. An oil reservoir.

d. Oil coolers and filter.

e. Pressure gauges, thermometers, a relief valve, an oil level indicator, all oil piping and valves.

Provisions must be made to warm up the heater drain pumps before placing them in operation, if this is required by the pump contractor.

Flow and pressure drop data used for level control valve sizing should be tabulated in the design criteria for the specific plant.

All level control valves that function during normal system operation must fail closed if there is a loss of control air. The bypass level control valve for the heater drain tanks must fail closed when there is a loss of control air to prevent uncontrolled drainage. All other bypass level control valves must fail open when there is a loss of control air. Hand wheels must be installed for the heater drain tank bypass level control valves to permit the manual operation of a valve that fails.

4.2.4   Moisture Separator-Reheater Drains and Vents

The turbine generator contractor must supply drain tanks or drain pots, control valves, bypass valves, and ms/rh safety relief valves necessary to drain and vent the moisture-separator reheater. The contractor's description of the ms/rh drain and vent system must be an integral part of the design criteria.

4.2.5   Gland Steam Condenser

The gland steam condenser is supplied by the turbogenerator manufacturer. It has connections on the channel side for condensate cooling and connections on the shell side for the seal steam inlet, and drains outlet. Normally, two exhaust blowers are mounted on the gland steam condenser to remove air.

4.2.6   Atmospheric Pressure Condensate Drain Tank

The atmospheric pressure condensate drain tank has connections toreceive drains from the gland steam condenser, condenser expansion joint seal overflow, vacuum breaker valve seal overflow, and the main feedpump outboard leakoff. Condensate level is controlled in the tank. The tank has connections for instrument taps, condensate drains, and an overflow.

The tank is constructed of carbon steel and designed to meet the requirements of ASME B&PV Code Section VIII, Division 1. The tank design pressure is 50 lb/in^2g. The design temperature is the saturation temperature at 50 lb/in^2g.

4.2.7   Valves

All valves are designed in accordance with ANSI B16.5.

Each heater drain pump is equipped with a combined minimum flow recirculation valve and check valve. The pump recirculation valve is an automatic modulating self-contained type.

All valves having their valve stems exposed to a vacuum must be equipped with lantern rings or bellows, to prevent air inleakage.

4.2.8   Piping

1. All piping in the heater drains and vents system is designed in accordance with ANSI B31.1, "Code for Pressure Piping." The design pressures and temperatures for the piping are discussed below:

a. Drain Piping

The design pressure and temperature of the drain piping located downstream of the feedwater heaters or drain tanks, up to and including the last isolation valve, is the same as that of the upstream feedwater heater or drain tank. Piping upstream of the feedwater heaters or drain tanks to the first upstream isolation valve has the same design pressure and temperature as that of the downstream feedwater heater shell or drain tank.

The piping located downstream of the bypass control valves and the APC tank level control valves (also the No. 7 heater drain control valves for a BWR) direct drains to the condenser and are subject to flashing flow. The design pressure for these lines is determined by calculating the maximum pressure in the line, assuming flashing. A larger than normal corrosion/erosion allowance must be added to the required pipe thickness to protect against flashing. The downstream piping is designed for a maximum fluid velocity of 100 feet per second at 100 percent load.

b. Vent Piping

Vent piping has the same design pressure and temperature as the shell side of the heater it vents, in the area between that heater and the last isolation valve. The design pressure of the vent piping located downstream of the isolation valve is the maximum operating pressure determined by flow calculations.

c. Drain Pump Discharge Piping

Drain pump discharge piping up to and including the level control valves and/or the last isolation valves, must have a minimum design pressure equal to the pump shutoff discharge pressure with maximum suction pressure.

Drain pump discharge piping coming from the level control valves or isolation valves to the connections in the condensate system has the same design pressure as the condensate piping to which it connects.

d. Drain Pump Suction Piping

Drain pump suction piping has the same design temperature
as the heater drain tank. Design pressure is the same
as that for the heater drain tank plus the static
pressure, difference between the tank at hi-hi level
and the pump suction.

2. Drain lines coming from the No. 3 and No. 7 (PWR) heater to
its respective heater drain tank must be sloped continuously
down to the tank and designed for self-venting flow. Drain
line connections to the drain tank should be located below
the normal operating level of the tank.

A flow element must be installed in the drain lines coming
from the No. 1, No. 2, No. 4, No. 5, and No. 6 feedwater
heaters, each reheater drain tank, and in the drain pump
discharge headers. A flow element must be installed in
the drain line coming from the No. 7 heater drain cooler
for a BWR. The location and design of the flow element
must prevent flashing inside the flow element.

Piping design must consider the limitations on the loading
of nozzles on pumps and feedwater heaters (these loadings
are subject to approval by the manufacturers).

The vent piping must have drains at all low points in the
system.

Vents are installed from each feedwater heater shell to the
condenser for removing noncondensable gases. An orifice
sized to accommodate one-half of one percent of the steam
flow entering each heater is located in each vent line.
For a PWR, a manual isolation valve is located in each vent
line. For a BWR, an automatic isolation valve is located
in each vent line.

Heaters with shell side relief valves are vented to the
condenser through a single line that is common to each
stage of feedwater heating. This vent line is sized
adequately to prevent one relief valve discharge from
obstructing the discharge of another relief valve. For
a BWR, the shell side relief valves of these heaters discharge to the condenser through this line. Separate vent
lines coming from each stage of feedwater heaters to the
condenser are installed to prevent the accumulation of
noncondensable gases in the lower pressure heaters.

Pockets or loop seals in the vent piping should be avoided to ensure that there is adequate venting between the heater and condenser during low differential pressure conditions or unit startup.

Heaters without shell side relief valves are vented to the condenser through separate heater vent lines.

A maintenance isolation valve is installed downstream from the level control valve located in the No. 2 feedwater heater drain going to the No. 3 heater drain tank. Manual isolation valves are installed upstream of all bypass level control valves that are routed to the condenser.

There must be sufficient static head available between the No. 1 and No. 2 heaters and between the No. 2 heaters and the bypass drain condenser connections to assure drainage during startup. For the Nos. 4, 5, 6, and 7 heaters, there must be sufficient static head available between the heater level and the condenser connection to ensure drainage during startup.

The No. 5 and 6 heaters must have power operated shell drain valves for removing moisture from the turbine extractions when the heaters are isolated. These drains shall be routed to the condenser.

The suction and discharge lines of each heater drain pump (HDP) are equipped with isolation valves. For a BWR, these isolation valves are motor operated.

Adequate pump suction head is made available by locating the heater drain tank at the highest possible elevation and designing the piping to maintain minimum NPSH requirements for the drain pumps during all operating conditions. Suction head is based on a static head available with the lowest operating drain tank level. The suction piping should include a startup strainer for each pump, which can be removed after the system is clean. Pressure taps upstream and downstream of the strainer should be included.

The heater drain pump suction piping is sized so that the hot water retention time is minimized, during depressuriza-tion of the drain tank (i.e., turbine trip), to limit or prevent flashing in the pump suction piping. The maximum velocity in the HDP suction piping is 15 ft/sec. The arrange-ment of the HDP suction piping should minimize turbulence and provide a flat velocity profile at the pump suction nozzle. For double-suction horizontal pumps, the suction piping should always be routed to the pump perpendicularly to the pump shaft and should have at least 5 diameters of straight length immediately upstream of the pump suction flange.

A corrosion allowance of 0.125 inches must be added to the minimum wall pipe thickness required for the design conditions for all flashing water drain piping downstream of the control valves to the point of discharge.

The drain level control valves are located as near to the point of discharge as possible. Tees with replaceable blind flanges are used between the control valves and the downstream vessel instead of elbows or bends.

The flashing water inlet connections to the heat exchangers are designed using the recommended method outlined in the Heat Exchanger Institute Standards on Feedwater Heaters.

Each heater drain pump has a minimum recirculation line piped to its corresponding heater drain tank and sized to meet the requirements for the manufacturer's recommended recirculation flow.

Check valves are installed in the normal drain lines routed from the ms/rh shell drain tank, unless this drain is used as a primary source of extraction steam.

Bypass level control valves are installed for alternate drains going from the reheater drain tanks to the condenser. The bypass level control valves and normal level control valves must have separate level controls from their respective drain tanks.

Check valves must be installed in the normal drain line routed from the reheater drain tanks to prevent reverse flow or flashing of the drains during transient operation. The check valve is located just downstream of the branch line tee that is routed to the reheater drain tank bypass level control valve. The piping between the reheater drain tank and the check valve should be of minimum length to reduce the volume of water flowing between the tank and the check valve which is subject to flashing during transient conditions. However, the vertical distance between the normal level in the reheater drain tank and the check valve should, if possible, supply enough static head to prevent cavitation through the check valve during maximum flow occurring under normal operating conditions.

Continuous venting of the reheaters is accomplished through the use of orificed vent lines (containing an open valve downstream of the orifice) routed to their respective feedwater heaters. The vent lines are connected to their respective reheater drain headers near the feedwater heaters. Reheater drain tanks are vented to their reheaters.

Startup venting and special venting requirements are fulfilled according to the turbine contractors recommendations.

The turbine contractor will recommend piping and valve sizes, since he is responsible for supplying the criteria for the design of the ms/rh drains.

The atmospheric pressure condensate drain tank must have a level control valve for normal operation and a bypass level control valve with the piping from both level control valves going to the condenser. In addition, the tank has a dump valve to permit tank drainage when the main condenser is not available. An overflow line is also provided.

The exhausters that remove air from the gland steam condenser discharge to a manifold that vents through the turbine building HVAC system or through a separate vent stack. A check valve located in each exhauster supply prevents backflow through the standby exhauster. The discharge piping from each exhauster must be capable of draining condensate accumulation back to the atmospheric condensate drain tank. The condensate level in the gland steam condenser outlet is maintained by a loop seal.

The drain piping from the Nos. 1, 2, 4, 5, and 6 heaters must have sampling connections located upstream of the level control valves. The drain piping from the No. 3 and No. 7 heater drain tanks must have sampling connections upstream of the bypass level control valves and heater drain pumps to ensure that water quality requirements are met before the drains are pumped forward.

4.2.9   Instrumentation

In addition to the instrumentation necessary to perform control and measurement functions, additional instrumentation must be available to measure system temperatures and pressures that are monitored by the BOP computer system, provide test information as required under ASME PTC-6, or supply additional information relating to the operation of system equipment. This instrumentation must include the following:

a. Temperature measurements in the suction of each pump.

b. Pressure measurements in the suction and discharge of each pump.

c. Temperature measurements in the drain lines routed from each feedwater heater, drain tank, and mfpt condenser.

d. Temperature measurements in the vent piping routed from the feedwater heater channel side relief valves.

e. Flow measurements in the No. 3 and No. 7 (PWR) heater drain pumps discharge piping.

If the data transmitted by the sensing elements is not needed for plant computer or logic functions, the information display on local panels will be sufficient.

Control room indication is required for the following:

a. Current to each drain pump motor.

b. Drain pump discharge pressure.

c. Condensate levels in all heaters and tanks equipped with level control.

d. High condensate levels present inside of heaters that are not equipped with level control.

e. Status of all automatic operating control and isolation valves. In addition, alarms are activated in the main control room if bypass level control valves open.

f. Low condensate level in heater drain tanks.

g. Opening of any orifice vent bypass valve.

All pressure monitoring connections must be equipped to facilitate on-line instrument calibration without disturbing station instrumentation.

All bypass level control valves have level controls that are separate from their respective normal level control valves. The level controllers have separate instrument piping routed from the controllers to the instrument connections on the heater or tank.

All heaters and heater drain tanks that are equipped with level instrument connections have a gauge glass that covers the entire fluid level range of the vessel.

4.2.10 Insulation

All piping and equipment located outside the condenser, where the temperature may exceed 150°F, must be insulated to prevent surface temperatures from exceeding 125°F. The extraction piping and heaters located inside the condenser have metallic shields to limit the heat loss to the condenser.

4.2.11 Steam Jet Air Ejector Intercondensers (BWR Only)

Each steam jet air ejector (sjae) intercondenser must have two drain paths routed to the main condenser hotwell. One path supplies normal operation drainage through a "fail open" drain control valve and a loop seal, while the other path must supply startup drainage through an impulse trap. The normally operating drains from both sjae's are "headered" downstream of the drain control valve and routed to the condenser through a common loop seal. The startup drains routed from both sjae's are "headered" downstream of all valves and directed to the condenser through a common line.

The shutoff valve in the normal drain line automatically opens when the pressure in the steam supply line routed to its respective sjae exceeds a set point that indicates startup. The valve must close automatically if its associated sjae is isolated. Each valve should have main control room position indication and manual control from the main control room.

The normal drain line must have an automatic valve located downstream from the loop seal to ensure that proper drain control is maintained. Instrumentation must be installed to detect a loss of water level in the loop seal (i.e., indicating the potential for steam blowby to the main condenser), and initiate automatic closure of this valve. With the valve closed, condensate draining from the active sjae will refill the loop seal and permit the valve to reopen. Manual control, valve position indication, and annunciation of valve closure must be located in the control room.

The startup drain must have an impulse trap (with strainer) mounted in the piping coming from the sjae intercondenser. A manual valve, check valve, and a normally open shutoff valve is also needed.

4.2.12 Off-Gas System

This system is applicable only to a BWR.

a. Off-Gas System Condenser

The off-gas system (OGS) condenser shell condensate level
is controlled by a normal LCV and a bypass LCV.

Both control valves discharge to the main condenser, and
valve control is accomplished by independent control trains
coming from shell-mounted level transmitters. Both valves
have control room position indication and high-low alarms.
The OGS condenser channel has a drain line and a pressure
relief valve that discharge to the main condenser. The
drain line has a manual, normally closed valve.

b. Off-Gas System Preheater

Condensate from the tube side of the OGS preheater is
drained to the main condenser by parallel lines, each
line equipped with a strainer and impulse trap.

Preheater channel pressure relief valves must discharge to
the main condenser, and each relief valve is bypassed by a
normally closed, manual valve that has a startup venting
capability.

4.2.13 Building Heat Exchangers and Absorption Chillers

The condensate from each absorption chiller drains to a separate
level control reservoir, and then to the condenser through a
level control valve (LCV). A high fluid level in the reservoir
must be alarmed in the main control room.

The condensate from the building heat exchangers drains to a
common level control reservoir, and is routed from the reservoir
to either unit's condenser through an isolation valve and a LCV.
If both condensers are out of service, the condensate is routed
through two pumps, a LCV, and an isolation valve to one of the
two auxiliary deaerators.

The controls for the isolation valves to the unit condensers
assures that drainage is directed to the same unit supplying
steam to the heat exchangers. The isolation valve control to
the auxiliary deaerators must be "remote-manual" and the primary
pump is manually started. The backup pump starts automatically
when there is a high fluid level in the reservoir if the pump is
in the automatic mode. Both pumps can only be stopped manually
by the operator.

131

The LCV's to both the unit condensers and the auxiliary deaerator are installed with the controls in parallel from the same level controller. A high level in the reservoir is signaled by an alarm in the control room.

The level control reservoirs are vented to their respective heat exchangers for the purpose of maintaining the same pressure in the reservoir and the heat exchanger. In addition, the building heat exchangers level control reservoir and each absorption chiller level control reservoir have orificed vents routed to the main condenser.

4.3 LOGIC FOR OPERATION

4.3.1 Heater Drain Tanks and Pumps

The heater drain tank pumps (HDP) are not needed for startup operation until 40-50 percent load is achieved. The drain tanks are drained through their respective bypass level control valves to the condenser. The heater drain flow bypass is required to clean the condensate during startup or when it is necessary to satisfy feedwater quality requirements during power operation. A handwheel is installed on each bypass level control valve so that it can be used in case the bypass valve operator fails to control the heater drain tank level.

Startup of any HDP may be controlled from a local panel or the main control room.

Two of the No. 3 HDP's (and one of the No. 7 HDP's for a PWR) are started at 40-50 percent turbine load and when the condensate quality, as determined by sampling from the HDT bypass drain line, is acceptable. The third No. 3 HDP is started at approximately 60 percent load. For a PWR, the second No. 7 HDP is started when approximately 50 percent load is achieved. Prior to reaching these load setpoints, the last HDP may be started manually to replace a tripped pump.

4.3.1.1 The following permissive interlocks must be satisfied before a HDP starts:

a. Adequate water level in the respective HDTs.

b. Bearing oil pressure greater than the low-low oil pressure setpoint (i.e., auxiliary lube oil pumps operating for at least five seconds). This is not applicable to vertical pumps with water lubricated bearings.

4.3.1.2 Fluid level in the No. 3 HDT (No. 7 for a PWR) is controlled by the heater drain tank normal LCV(s) located in the HDP discharge line routed to the condensate system or by the tank bypass level control valve routed to the condenser. The pertinent fluid levels in the heater drain tank during operation, based on a tank diameter of 10 feet, are as follows:

a. Low level - 4 feet below tank centerline.

b. Normal level range - 6 inches to one foot above tank centerline.

c. Bypass level range - 18" to 24" above tank centerline.

d. Hi level - 3 feet above tank centerline.

e. Hi-hi level - 4 feet above tank centerline.

The system responses to each of these fluid levels in the heater drain tanks are given below:

f. A low level alarm is actuated in the main control room and, after a 10-second delay, trips all drain pumps. The pumps must be restarted manually by the operator.

g. If a drain pump is tripped, a normal level control valve fails, or there is excessive heater tube leakage, the tank bypass level control valve opens when the fluid level in the tank rises to the bypass level range. A limit switch on the bypass level control valve actuates a continuous display that indicates an unseated valve.

h. If the fluid level continues to rise above the heater bypass level to the hi level, a high-level alarm sounds in the main control room alerting the operator.

i. If the level proceeds to rise above the hi level to the hi-hi level, the normal No. 2 heater drain level control valve and the ms/rh shell drain level control valve close, after a 5-second delay, causing the No. 2 heater drains and the ms/rh shell drains to be bypassed to the condenser.

133

4.3.1.3 All of the heater drain pumps are tripped because of the following:

a. Low-HDT water level.

b. Low-low bearing oil pressure. (Low oil pressure actuates an alarm and signals the auxiliary lube oil pump to start.)

c. Main turbine trip.

A signal indicating a low differential pressure existing across any one of the three No. 3 heater drain pumps causes one of the two normal level control valves to close. This prevents a pump runout condition from occurring. A pump runout condition could trip the remaining No. 3 heater drain pumps because of a motor overload (i.e., over-current). Prevention of a runout condition also reduces the potential for pump low npsh problems.

4.3.1.4 PWR Only--If the level in the No. 7 heater drain tank rises from the hi level to the hi-hi level, the normal No. 6 heater drain level control valve closes, after a 5-second delay, causing the No. 6 heater drains to be bypassed to the condenser.

If the three hotwell pumps and three condensate booster pumps can handle 100 percent of the bypass flow during 100 percent load, the bypass level control valve is designed to accommodate the total bypass drain flow, and no runout protection for the No. 7 drain pumps is available. However, if the hotwell and booster pumps cannot handle 100 percent of the bypass drain flow from the No. 7 heater drain tank, the bypass level control valve is sized to bypass the maximum drain flow that the condensate system can accommodate with three hotwell pumps and three condensate booster pumps operating.

The discharge from the No. 7 drain pumps is controlled by a level control valve that maintains the No. 7 heater drain tank level. The control valve is sized to accommodate all of the drain flow during 50 percent guaranteed heat balance load with one pump operating, and all of the drain flow during 105 percent guaranteed heat balance load with two pumps operating. If required to prevent a HDP runout and trip, the level control valve's opening must be limited to 50 percent capacity whenever at least one pump is out of service.

4.3.2   Feedwater Heaters Nos. 1, 2, 3, 4, 5 and 6

Fluid level in these feedwater heaters is controlled by a normal level control valve or the bypass level control valve. The governing criteria for fluid levels inside the feedwater heaters during operation are as follows:

a. Normal level range - To be specified by feedwater heater contractor.

b. Bypass level range - To be specified by feedwater heater contractor.

c. Hi level - The feedwater heater centerline.

d. Hi-hi level - One foot above feedwater heater centerline.

e. Lo level - To be specified by feedwater heater contractor.

The fluid level at which the normal level control valve is fully closed (i.e., the lower limit of the normal level control valve control band) is specified by the heater contractor, and is the minimum feedwater heater shell level required to ensure proper operation of the drain cooler portion of the feedwater heater. The fluid level for which the bypass level control valve is fully open (i.e., the upper limit of the bypass level control valve control band) should be below the anti-flash baffles (if provided) in low pressure heaters (No. 5 and No. 6). For the heaters with drain coolers (heater Nos. 1, 2, and 4) the level at which the bypass LCV is fully open may be above the bottom tubes, if that prevents flashing in the draincooler inlet during abnormal operation (i.e., with 150 percent of normal drains flow).

A rise of fluid to the bypass level in the feedwater heater causes the bypass level control valve to open. A limit switch located on the bypass level control valve actuates a continuous display in the main control room, indicating an unseated valve.

A hi level in the feedwater heaters actuates an alarm in the main control room, alerting the operator.

A rise in a feedwater heater's fluid level to the hi-hi level causes it and its affected stream of heaters to become isolated. Heater isolation trips the power operated extraction nonreturn valves, closes the motor operated isolation valves in the extraction lines routed to their respective heaters (Nos. 1, 2, and 4 only), and closes the motor operated isolation valves located upstream and downstream from the heater channels. Also, all drains and vents coming from the reheaters or higher pressure heaters are isolated when a feedwater heater's fluid level reaches the hi-hi level.

A low level in a heater actuates an alarm in the main control room.

Isolation of a stream of Nos. 5, 6, and 7 heaters causes the drains from the No. 4 heaters to be bypassed to the condenser. It also causes the opening of the power operated Nos. 5 and 6 heater shell drain valves. The isolation of the channel side of the heaters in conjunction with the opening of bypass LCV and heater shell drain valves should reduce the heater water level and prevent turbine water induction. In addition, a hi-hi level in the No. 7 heater drain tank causes the No. 6 heater LCV to close causing the drains from the No. 6 heater stream to be bypassed to the condenser.

4.3.3   The No. 3 Feedwater Heaters

The No. 3 heater is designed as a "dry" heater that has no fluid level control. If a large heater tube leak or other abnormality causes water to rise one foot above the heater centerline in any No. 3 heater, the affected No. 3 and No. 4 heater stream (i.e., a, b, or c) must be isolated. If the No. 3 heater drain tank overflows into the No. 3 heaters, it will cause all three streams of heaters to become isolated and the reactor to trip because of a loss of feedwater. Since the No. 3 heater is designed as a "dry" heater with no valves located between the heater and the No. 3 heater drain tank, there are no other fluid levels inside the heater that cause a system action.

4.3.4   No. 7 Feedwater Heaters

PWR Only--The No. 7 heater is designed as a "dry" heater and there is no fluid level control for the drains routed to the No. 7 heater drain tank.

BWR Only--The No. 7 heater is designed as a "dry" heater and there is no fluid level control for the drains routed to the No. 7 drain cooler.

4.3.5 Moisture Separator and Reheaters

The moisture-separator drains, first-stage reheater drains, and the second-stage reheater drains are controlled according to the turbine manufacturer's recommendations.

The first and second stage reheater vents are also controlled according to the turbine manufacturer's recommendations.

4.3.6 Gland Steam Condenser

The gland steam condenser drains to the atmospheric drain tank through a loop seal and does not have a controlled condensate level. If the condensate level inside the loop seal reaches a high level, a fluid level switch activates a control room alarm.

The gland steam condenser vacuum data is displayed and logged in the control room. A low vacuum signal will alarm and issue a delayed start signal to the standby exhauster.

The exhausters alternately start automatically, but must also have the capability of manual operation from the control room.

4.4 MAINTENANCE AND INSPECTION CRITERIA

4.4.1 Heater Drain and Vents System Maintenance and Inspection

Since no component of the heater drains and vents system has a safety class, inservice inspection as defined in ASME Boiler and Pressure Vessel Code, Section XI, is not required. However, design of the system must provide for periodic functional tests and visual inspections that are performed in conjunction with unit outages.

Design of the heater drains and vents system must include a means for component isolation permitting on-line maintenance of the following equipment:

1. Heater drain pump.

2. Any stream of feedwater heaters not located in the condenser neck (except No. 3 feedwater heater).

On a BWR only, shielding must be installed where needed.

4.4.2   The design of the turbine building must allow for the maintenance of all system pumps, feedwater heaters, reheaters, and LCVs. Access ways must be provided to permit the movement of equipment into or out of key areas. Lifting devices must be available for handling large components. Overhead space is required to lift heater drain pump impellers.

All equipment must have drain and vent connections to remove water from equipment internals.

4.4.3   BWR Only--Facilities must be available inside the turbine building to decontaminate equipment that is to be returned to the factory for maintenance. DG-M12.2.1, "CHEMICAL TREATMENT--Tool and Equipment Decontamination Facilities" must be consulted for details.

4.5   SYSTEM INTERFACES

The following documents will assist in defining system interfaces and provide additional insight into the design of the heater drains and vents system for a 3800 megawatt-thermal nuclear plant having light-water reactors.

a.   EN DES Design Guide DG-M2.2.1, "TURBINE GENERATOR HEAT CYCLE-- Main Feedwater System," Tennessee Valley Authority: Knoxville.

b.   EN DES Design Guide DG-M2.3.1, "TURBINE GENERATOR HEAT CYCLE-- Condensate System."

c.   EN DES Design Guide DG-M2.4.1, "TURBINE GENERATOR HEAT CYCLE-- Extraction Steam System," (to be issued).

d.   EN DES Design Guide DG-M2.7.1, "TURBINE GENERATOR HEAT CYCLE-- Condenser and Auxiliaries System," (to be issued).

e.   EN DES Design Guide DG-M2.8.5, "TURBINE GENERATOR HEAT CYCLE-- Pressure Drop Calculations for Steam and Condensate Piping."

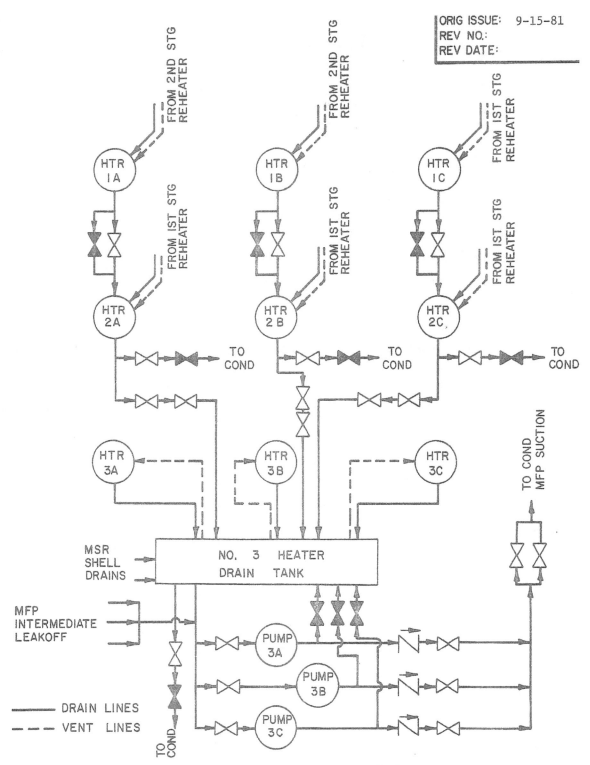

HIGH PRESSURE HEATER DRAINS AND VENTS

139

ORIG ISSUE: 9-15-81
REV NO.:
REV DATE:

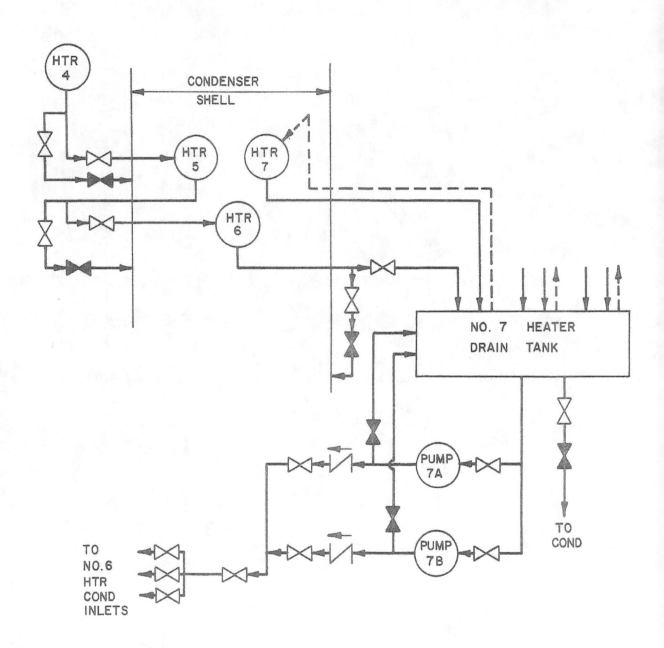

LOW PRESSURE HEATER DRAINS AND VENTS FOR A PRESSURIZED WATER REACTOR

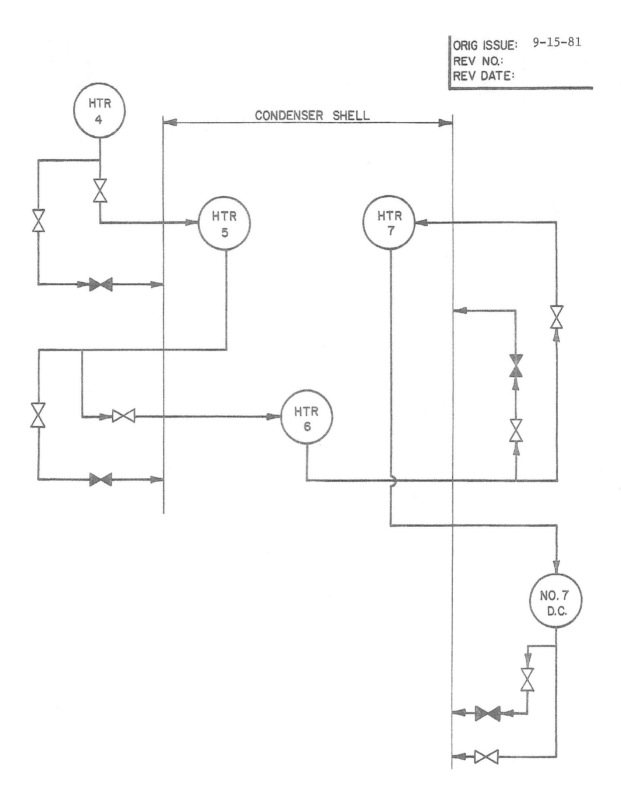

ORIG ISSUE: 9-15-81
REV NO.:
REV DATE:

CONDENSER SHELL

LOW PRESSURE HEATER DRAINS AND VENTS FOR A BOILING WATER REACTOR

ORIG ISSUE: 9-15-81
REV NO.:
REV DATE:

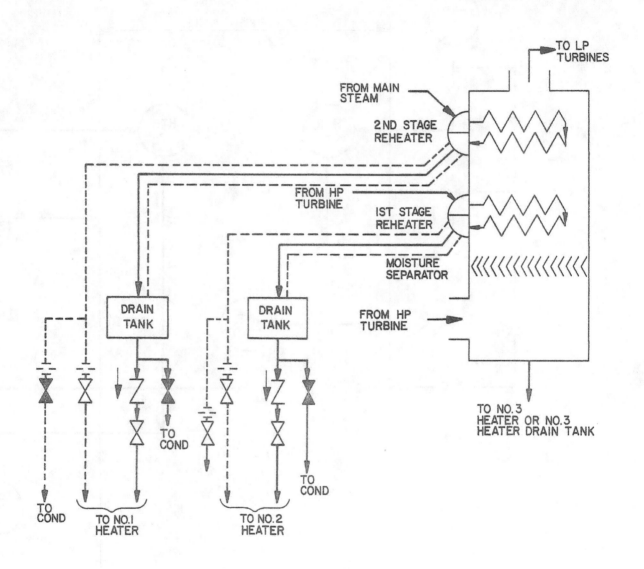

MOISTURE SEPARATOR REHEATER DRAINS AND VENTS

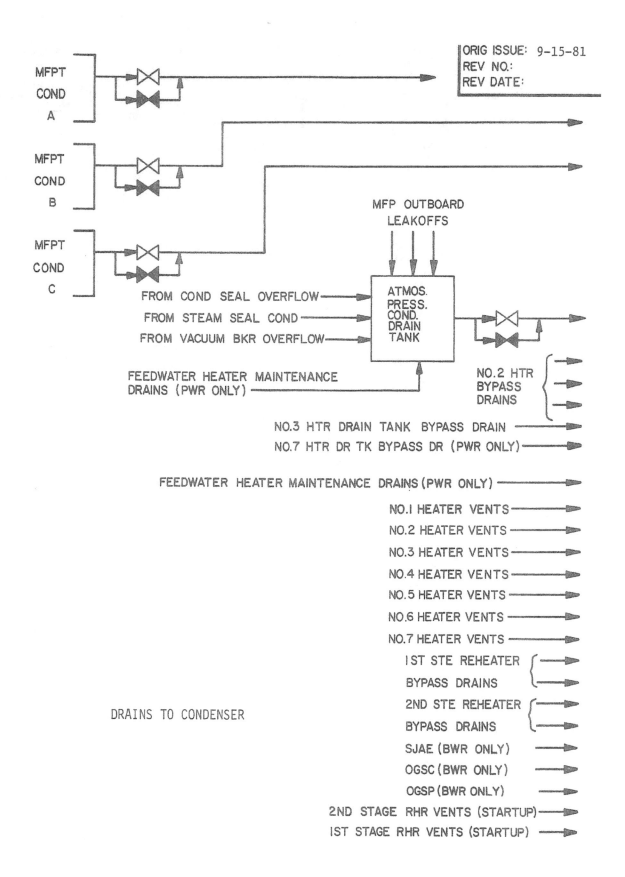

ORIG ISSUE: 9-15-81
REV NO.:
REV DATE:

MFPT COND A

MFPT COND B

MFPT COND C

MFP OUTBOARD LEAKOFFS

FROM COND SEAL OVERFLOW

FROM STEAM SEAL COND

FROM VACUUM BKR OVERFLOW

ATMOS. PRESS. COND. DRAIN TANK

FEEDWATER HEATER MAINTENANCE DRAINS (PWR ONLY)

NO.2 HTR BYPASS DRAINS

NO.3 HTR DRAIN TANK BYPASS DRAIN

NO.7 HTR DR TK BYPASS DR (PWR ONLY)

FEEDWATER HEATER MAINTENANCE DRAINS (PWR ONLY)

NO.1 HEATER VENTS

NO.2 HEATER VENTS

NO.3 HEATER VENTS

NO.4 HEATER VENTS

NO.5 HEATER VENTS

NO.6 HEATER VENTS

NO.7 HEATER VENTS

1ST STE REHEATER BYPASS DRAINS

2ND STE REHEATER BYPASS DRAINS

SJAE (BWR ONLY)

OGSC (BWR ONLY)

OGSP (BWR ONLY)

2ND STAGE RHR VENTS (STARTUP)

1ST STAGE RHR VENTS (STARTUP)

DRAINS TO CONDENSER

143

ORIG ISSUE: 9-15-81
REV NO.:
REV DATE:

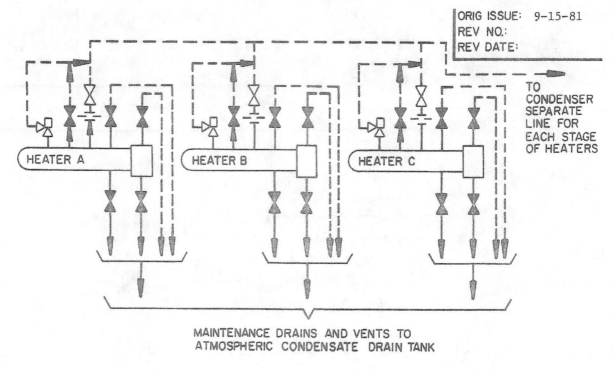

TO
CONDENSER
SEPARATE
LINE FOR
EACH STAGE
OF HEATERS

MAINTENANCE DRAINS AND VENTS TO
ATMOSPHERIC CONDENSATE DRAIN TANK

HEATERS NO.1, NO.2, NO.3, & NO.4

HEATERS NO.5, NO.6, & NO.7

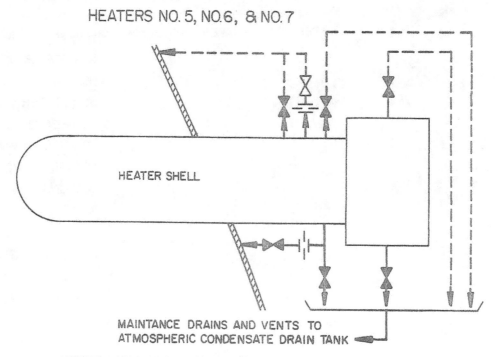

HEATER SHELL

MAINTANCE DRAINS AND VENTS TO
ATMOSPHERIC CONDENSATE DRAIN TANK

HEATER SHELL VENT DETAILS FOR A PRESSURIZED WATER REACTOR

ORIG ISSUE: 9-15-81
REV NO.:
REV DATE:

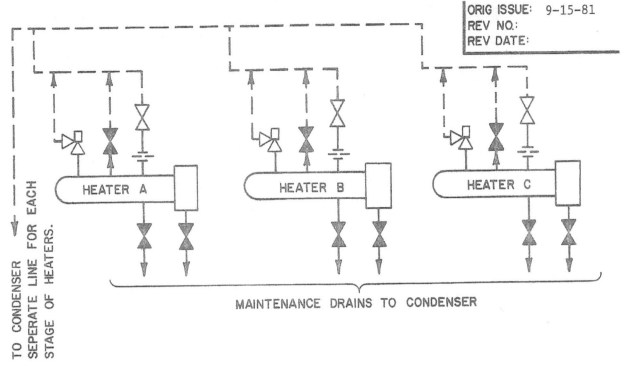

TO CONDENSER SEPERATE LINE FOR EACH STAGE OF HEATERS.

MAINTENANCE DRAINS TO CONDENSER

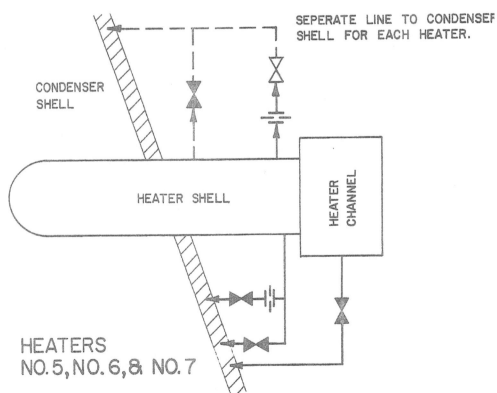

SEPERATE LINE TO CONDENSER SHELL FOR EACH HEATER.

CONDENSER SHELL

HEATER SHELL

HEATER CHANNEL

HEATERS NO.5, NO.6,& NO.7

HEATER SHELL VENT DETAILS FOR A PRESSURIZED WATER REACTOR

APPENDIX B - DESIGN CALCULATIONS FOR THE HEATER
DRAINS AND VENTS SYSTEM

B.1   GENERAL

This design guide attempts to outline the calculations that are needed
for the basic design of the heater drains and vents system. In addition
to aiding the engineer responsible for the system design, this design
guide presents a format for performing these calculations. This ensures
that the system design is consistent with the designs of other TVA plants
and that the calculations are a convenient source of reference.

B.2   DESIGN INFORMATION

The information described below is needed to perform the heater drain
system calculations.

2.1   Turbine Generator Contractor Information

Turbine generator heat balances at 105 percent, 100 percent,
75 percent, 50 percent, and 25 percent of the NSSS rating.

2.2   Plant Layout and Piping Group Information

Piping isometric sketches show the lengths between changes in
pipe diameter, branches in piping and the approximate location
of valves and fittings. The isometric should show the relative
centerline elevations of the feedwater heaters, drain tanks,
control valves, and drain coolers. The plant layout and equipment group should refer to the design criteria diagrams for piping
requirements.

2.3   Heater Drain Pump Contractor Information

Head capacity curves which show pump characteristics for flows
between pump shutoff and 150 percent of the specified pump flow.
Minimum recommended flow and minimum recommended npsh requirements
should be shown on the curves. For preliminary calculations,
estimated curves should be prepared based on similar pumps from
several manufacturers.

B.3   HEATER DRAIN PUMP CALCULATION SHEET

A Heater Drain Pump Calculation Sheet is included as Table 1.

For a PWR, one calculation sheet should be used for the No. 3 heater drain pump and another sheet for the No. 7 heater drain pump.

For a BWR, only one calculation sheet for the No. 3 heater drain pump is needed. The percentage columns on the calculation sheet refer to system design points that are determined by turbine load. The blank spaces in each column are for tabulating calculation results at these design points. The calculations are discussed in the sections that follow.

B.4   DRAIN PIPING AND FITTING PRESSURE DROP

Mechanical Design Guide DG-M2.8.5 (see section 4.5.e), should be used for making these calculations. The heater drain lines should be divided into two parts, one upstream and one downstream from the level control valve. For pumped drains, the lines should be divided into three parts, one upstream of the heater drain pumps, one downstream from the heater drain pumps to the level control valve, and one downstream of the level control valve. For pumped drains, the pressure at the entrance to the condensate system is obtained from the Condensate System Calculation Sheet (see section 4.5.b). Confirm that the condensate system pressures were calculated by assuming the maximum head rise of the condensate system pumps, between the design point and shutoff (see section B.5.1.1 of DG-M2.3.1, "Condensate System"). The remaining pressures, flows, temperatures, and enthalpies for all drains should be obtained from the turbine-generator heat balance.

Piping and fitting pressure drops should be calculated for 100 percent load and the remaining load points can be estimated by using the ratio of the head losses to the square of the flows, i.e.:

$$H_x = H_{100}(Q_x/Q_{100})^2$$

where:

H_x = Pressure drop, lb/in^2 (load X; X = 25%, 35%, 50%, 75%, and 105% load)

H_{100} = Pressure drop, lb/in^2 (100% load)

Q_x = Flow at load point X, lb/hr

Q_{100} = Flow at 100% load, lb/hr.

The calculations should account for pipe friction losses and head losses in fittings, flow nozzles, and valves. Pipe sizes should be chosen to limit velocities to a maximum of 20 ft/sec., except for piping passing two phase flow (see Section 4.2.9, item b. of part 1, and HDP suction piping, Section 4.2.8, part 2 of this design guide). When calculating friction loss for hotwell and condensate booster pump suction and discharge piping, assume schedule 40 pipe for pipe sizes 10 inches and larger.

Piping isometric sketches from the plant layout and piping group should be used to obtain piping lengths, the number of pipe fittings, etc.

B.5   STATIC HEAD PRESSURE DIFFERENCES

Differences in elevation should be converted to their corresponding static head pressure differences, measured in lb/in^2, as follows:

$$lb/in^2 = Feet/(144 \times SV)$$

where:

> SV = Specific volume of water in ft^3/lb at saturation temperature, taken from ASME steam tables, Table 2.

B.6   DRAIN PUMP FLOWS

The pump flow for each of the heater drain pumps can be calculated as follows:

$$G = (W \times SV \times 7.48) / (60 \times N)$$

where:

> G = Pump flow, for each pump, gal/min

> W = Total heater drain flow, taken from the turbine generator heat balance, lb/hr.

> SV = Specific volume of water in ft^3/lb taken from ASME Steam Tables, Table 2, for the temperature shown on the heat balance.

> N = Number of drain pumps operating for a particular calculation.

B.7   CALCULATING DRAIN PUMP HEAD AND FLOW

7.1   For preliminary pump sizing calculations, a 25 lb/in^2 differential pressure should be assumed across the heater drain pump discharge control valves for a flow occurring at 100 percent load. For other flows, the differential pressure must be prorated based on the methods presented in section 4.2.3 of this design guide.

7.2   Calculations that determine HDP head and flow requirements should be made using Table 1 of this appendix.

7.3   The HDP discharge pressure should be calculated by starting with the condensate system pressure and working upstream. The HDP suction pressure should be calculated by starting with the HDT pressure and working downstream.

7.4   The HDP head required at each load should be calculated using Table 1. A separate calculation should be made for each operating requirement described in section 4.1.3 of this design guide. The head required, including 5 percent margin, should be represented graphically as in Figure B-1.

B.8   CALCULATING HEATER DRAIN PUMP DESIGN POINT

The design point for the HDP should be 1.05 × head required for 100 percent load, and governed by the conditions outlined in section 4.1.3, item a, of this design guide. This design point corresponds to the HDP Head Available during 100 percent load (see Table 1).

The HDP specification specifies that the shutoff head for the pump will be 120 to 130 percent of the total head existing for the design point. The smallest head rise from rated conditions to shutoff should be used for HDP sizing. After researching HDP bids and coordinating with manufacturers, a HDP characteristic curve is estimated, including the "droop" from the design point to the runout condition. A typical HDP characteristic curve is shown in Figure B-2.

Using the characteristic curve and design point, the "HDP Head Available" for each load condition can be established.

A family of HDP curves for one, two, and three pump operation can be constructed and superimposed on the "head required" curve.

149

ORIG ISSUE: 9-15-81
REV NO.:
REV DATE:

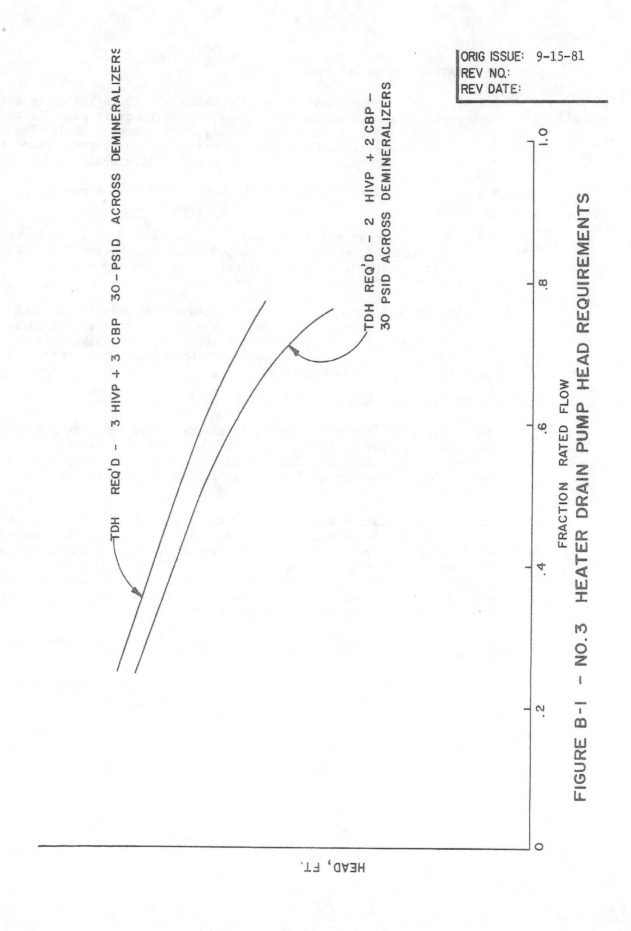

FIGURE B-1 - NO.3 HEATER DRAIN PUMP HEAD REQUIREMENTS

ORIG ISSUE: 9-15-81
REV NO.:
REV DATE:

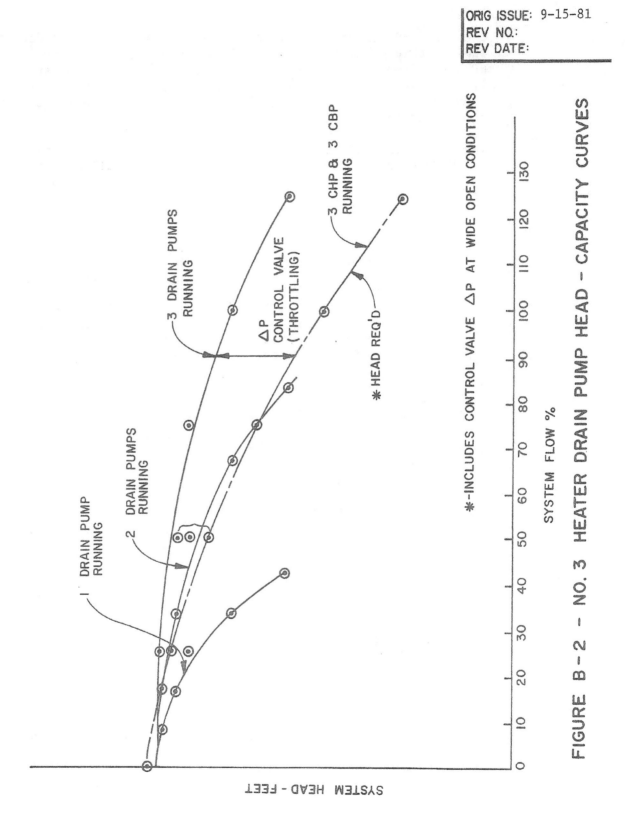

FIGURE B-2 - NO. 3 HEATER DRAIN PUMP HEAD - CAPACITY CURVES

151

Check the HDP characteristic curves established above to determine whether they meet all required performance criteria. Verify that the NPSH requirements for the HDP are satisfied for the operating range. If the characteristics selected are satisfactory, this indicates that they are sufficiently conservative so that pumps meeting the standard specifications will also meet all system requirements.

The available NPSH should be calculated using Table 1, and compared with the NPSH required by various manufacturers, for all load conditions, as a check for adequacy. The minimum NPSH available should be included in the standard specification.

B.9   SYSTEM CALCULATIONS USING ACTUAL PUMP CURVES

9.1   When the heater drain pumps are purchased, the condensate system calculations must be completed using the hotwell and condensate booster pump head-flow characteristic curves. The HDP Calculation Sheet (i.e., Table 1), the "HDP Head Required", and "HDP Head Available" should be revised. These calculations should be done using the same procedures described in section 8.0, but substituting the actual curves for estimated curves.

9.2   Table 1 should be completed to account (1) for three hotwell and three condensate booster pumps in service, and (2) for two hotwell and two condensate booster pumps in service. Calculations should verify that the HDP's meet the performance requirements outlined in section 4.1.3 of this design guide.

9.3   A separate calculation to determine actual system pressures should be made, and the control valve pressure drop (i.e., Table 1, line 9) should also be revised to reflect actual ΔP's dissipated by the control valve. This is accomplished by working backwards from line 13 to line 9. The ΔP's are used to size and procure the HDP control valves.

B.10   CONTROL VALVE PRESSURE DROPS AND FLOWS FOR CASCADED DRAINS

The turbine generator heat balance for 100 percent load is used to obtain the drain flows for heaters No. 1, No. 2, No. 4, No. 5, and No. 6 that are recorded in Table 2.

The flows shown for the heat balance should be divided by the number of heater streams to obtain the flow from each heater. These flows are entered in Table 2 as the normal flow for the control valve. The maximum control valve flow is 150 percent of the normal flow, and the

152

minimum control valve flow is 15 percent of the normal flow. The heater drain temperatures during normal flow and maximum flow are taken from the heat balance for 100 percent load. The heater drain temperatures at minimum flow are determined by extrapolating the temperature of the condensate entering the heater during 15 percent load from higher load heat balances. The extrapolated temperature should then be added to the design drain cooler approach temperature (i.e., 10°F).

The pressure in the heater shell is taken from the turbine generator heat balance during 100 percent load, for each feedwater heater. The piping pressure drop, with the appropriate adjustment for different flows, is subtracted from the heater shell pressure; the static head (in lb/in^2) is then added to obtain the control valve inlet pressure. The inlet pressures are entered into Table 2 for the normal and maximum flows. The inlet pressures during minimum flow are obtained in a similar manner except the heater shell pressure is assumed to be 15 percent of the pressure during 100 percent load.

The pressure downstream of the control valve (cv) is equal to the pressure in the downstream vessel plus the ΔP in the piping between the cv and the downstream vessel. This pressure drop calculation is based on flashing two phase flow, and static head can be ignored. Pressure drops are calculated and entered into Table 2 for both normal and maximum flows. The pressure drops at minimum flow are obtained in a similar manner except the heater shell (or drain tank) pressures are assumed to be 15 percent of the pressure during 100 percent load.

The actual ΔP across a valve is the difference between the inlet and outlet pressures calculated above.

The allowable pressure drop, C.V.ΔP_{all}, for sizing the control valve is calculated as follows[1]:

$$C.V.\Delta P_{all} = K_m (P_1 - r_c P_v)$$

where: K_m = valve recovery coefficient for the particular valve (Examples are given in Table 4)

P_1 = control valve inlet pressure, lb/in^{2}a

P_v = Vapor (saturation) pressure of the fluid at the temperature of the upstream heater

r_c = critical pressure ratio, as shown in Figure 1

---

1. "Valve Sizing for Cavitating and Flashing Liquids," Fisher Controls, August, 1967.

The control valve flow coefficient, C_v, can be estimated as follows:

$$C_v = \frac{G}{\sqrt{\Delta P - SV}}$$

C_v = Valve flow coefficient

ΔP = Smaller of the actual or allowable control valve pressure drop, lb/in^2.

SV = Specific Volume of water in ft^3/lbm, at saturation temperature taken from ASME Steam Tables, Table 2.

B.11 CONTROL VALVE PRESSURE DROPS AND FLOWS FOR PUMPED DRAINS

These calculations are performed after the purchase of the hotwell pumps, condensate booster pumps, and heater drain pumps; completion of the system calculations (see section 9.3, Appendix B). The control valve pressure drop is calculated for the conditions outlined in section 4.1.4 of this design guide.

The No. 3 HDP control valves are sized by calculating the C_v of each valve for the conditions outlined in section 4.1.2 of this design guide. Actual pressure drop across the valves is calculated using Table 1. $C.V. \Delta P_{all}$ is calculated using the equation outlined in section 10.0, Appendix B. Since the two control valves work together, the flow for each valve is 1/2 the flow through the HDPs. The parameters associated with rated flow conditions and minimum flow conditions should be included in the valve specification.

The maximum C_v for one No. 3 HDP control valve is calculated using the methods outlined in section 4.1.4, item b, of this design guide. The maximum allowable flow that prevents runout of the two HDPs is determined by the lowest of the following:

a. The flow associated with two HDPs running when the minimum NPSH available equals the NPSH required.

b. Twice the flow corresponding to a pump bhp equal to the HDP motor rating.

c. Twice the maximum flow shown on the vendor's pump curve.

The parameters associated with the maximum control valve size should be included in the specification.

The actual ΔP across the valve is the difference between the inlet and outlet pressures calculated above.

The allowable pressure drop used for sizing the control valve ($C.V.\Delta P_{all}$) is calculated (see section 10.0 of Appendix B) and compared to the actual control valve pressure drop. The control valve flow coefficient, C_v, is estimated also.

TABLE 1. HEATER DRAIN PUMP CALCULATION SHEET

Percent of NSSS Guaranteed Rating	105%	100%	75%	50%	35%
1. Total heater drain flow, lb/hr					
2. Flow per pump, lb/hr					
3. Spec. Vol at Heat Bal. temp, ft^3/lb					
4. Pump Flow per pump, gal/min					
5. Condensate Sys. Pressure, lb/in^{2}a					
6. Friction Loss - cv to Cond. Sys., lb/in^2					
7. Static Head - cv to Cond. Sys., lb/in^2					
8. Pressure at cv Exit, lb/in^{2}a					
9. Control Valve Pressure Drop, lb/in^2					
10. Pressure at cv Entrance, lb/in^{2}a					
11. Friction Loss - HDP to cv, lb/in^2					
12. Static Head - HDP to cv, lb/in^2					
13. HDP Discharge Pressure, lb/in^{2}a					
14. Heater Drain Tank Pressure, lb/in^{2}a					
15. Static Head - HDP to Htr. Dn. Tank, lb/in^2					
16. Friction Loss - HDP Suction Piping, lb/in^2					
17. HDP Suction Pressure, lb/in^{2}a					
18. HDP Head Required, lb/in^{2}a					
19. HDP Head Required, ft					
20. HDP Head Available from pump curve, ft					
21. HDP Head Available, lb/in^{2}a					
22. HDT Saturation Pressure, lb/in^{2}a					
23. HDP npsh Available, lb/in^2 (17-22)					

156

TABLE 2

Control Valve Sizing Calculations for Level
Control valves Used During Normal Operation

Heater Flow, lb/hr	Temp.	Friction ΔP, lb/in^{2}a	Static ΔP, lb/in^{2}a	cv Inlet Press, lb/in^{2}a	C.V.ΔP, lb/in^2	C.V.ΔP Allowable, lb/in^2
1(normal & bypass)	Max. Normal Min.					
2	Max. Normal Min.					
4	Max. Normal Min.					
5 (normal bypass)	Max. Normal Min.					
6	Max. Normal Min.					
7 (BWR)	Max. Normal Min.					
APC Drain Tank	Max. Normal Min.					

TABLE 3

Control Valve Sizing Calculations
for Bypass Level Control Valves

Heater Flow, lb/hr	Temp.	Friction ΔP, lb/in^{2}a	Static ΔP, lb/in^{2}a	cv Inlet Press, lb/in^{2}a	C.V.ΔP, lb/in^{2}a	C.V.ΔP_{all}, lb/in^{2}a
2	Max. Normal Min.					
4	Max. Normal Min.					
6	Max. Normal Min.					
7 (BWR)	Max. Normal Min.					
No. 3 Htr Dr.	Max. Normal Min.					
No. 7 Htr DR (PWR)	Max. Normal Min.					
APC Tank	Max. Normal Min.					

158

Chapter 6
Design Guide DG-M2.7.1: Condenser and Auxiliaries

1.0 GENERAL

This design guide is an aid for the design of the non-safety-related condenser and auxiliaries of a 3800 MWT nuclear plant with a light water cooled reactor.

Use of this design guide helps the engineer prepare a draft design criteria for a specific project. An engineer with a thorough understanding of the system should prepare the design criteria and perform the design calculations.

The condenser and auxiliaries have no safety related function; therefore, the discussion of such functions is not within the scope of this document.

1.1 ABBREVIATIONS

BWR	-	Boiling water reactor
HP	-	High pressure
in. HgA	-	Inches of mercury absolute
IP	-	Intermediate pressure
LP	-	Low pressure
MCR	-	Main control room
NRC	-	Nuclear Regulatory Commission
NSSS	-	Nuclear steam supply system
ppb	-	Parts per billion
PWR	-	Pressurized water reactor
scfm	-	Standard cubic feet per minute
SJAE	-	Steam jet air ejector

1.2 APPLICABLE CODES AND STANDARDS

The following standards shall apply to the design, construction, and performance of the equipment described in this document.

1.2.1 Condenser System

Heat Exchange Institute Standards for Steam Surface Condensers

1.2.2 Condenser Vacuum System

a. ASME Boiler and Pressure Vessel Code, Section VIII, Division 1, "Pressure Vessels"
b. ANSI B31.1, "Code for Pressure Piping--Power Piping"
c. Heat Exchange Institute Standards for Steam Jet Air Ejectors
d. Hydraulic Institute Standards
e. TEMA Class C

2.0 FUNCTIONAL DESCRIPTION

   2.1 CONDENSER SYSTEM

      The condenser system condenses steam from the main turbine and
      feed pump turbines, and is used to establish and maintain the
      vacuum required for operation of the turbines. It also acts
      as a heat sink for the turbine bypass system and for miscellaneous
      drains, vents, and steam dumps. The system deaerates the
      secondary cycle condensate and stores it to meet the condensate
      system requirements.

      The major components of the condenser system are the main
      condenser and feed pump turbine condensers. The system
      encompasses those regions enclosed by the main condenser and
      the feed pump turbine condensers, including all connections,
      appurtenances, and interconnecting piping.

      The main condenser is located directly below the LP turbines
      and connects directly to the turbine exhaust necks. Turbine
      exhaust steam flow is vertically downward and transversely
      across the tubes. A number of low pressure horizontal feed-
      water heaters are located in the neck of the main condenser.

      The total condensate formed, including that from miscellaneous
      drains, vents, and steam dumps is deaerated before flowing
      from the hotwell to the hotwell pumps.

      The main feed pump turbine condensers are located directly
      below and connect to the feed pump turbine exhaust necks.
      Turbine exhaust flow is vertically downward across the tubes.
      The condensate formed is drained to the main condenser hotwell.

      Proper condenser system vacuum is maintained by the collection
      of non-condensible gases in the main condenser and their removal
      by the condenser vacuum system. The feed pump turbine condensers
      are vented to the low pressure zone of the main condenser.

      Cooling water is provided by the condenser circulating water
      system. The heat transferred to the cooling water is ultimately
      released to the atmosphere by cooling towers. A continuous
      tube cleaning system is supplied for the main condenser.

   2.2 CONDENSER VACUUM SYSTEM

      The condenser vacuum system removes all noncondensible gases
      (including air and any radiolytic products originating in the
      reactor) from the condensers to initially evacuate the condenser
      system and maintain proper condenser vacuum during operation. The
      system also minimizes any radioactive release to the environment.

The condenser vacuum system is located in the turbine building. The system's boundaries extend from the connections for the main condenser and feed pump turbine condenser air take-off to the inlet of the turbine building ventilation system. (For a BWR unit, they extend to the interface with the reactor off-gas system.)

Lines from the main condenser air take-off connections (one connection for each tube bundle) connect to the air removal equipment suction header, as do the feed pump turbine condenser vent lines.

2.2.1   PWR Only--The air removal equipment consists of three half-capacity parallel mechanical vacuum pumps located as near the main condenser as is practical. Each pump is equipped with normally open maintenance isolation valves and a discharge check valve.

The vacuum pumps discharge into a common header leading to the atmospheric cleanup equipment. A safety valve in this header protects the downstream components from over-pressurization.

The atmospheric cleanup equipment, consisting of an air heater and a carbon adsorber unit, has manually operated, normally open, maintenance isolation valves at the entrance and discharge. A bypass containing an automated, normally closed isolation valve is installed around the cleanup equipment and isolation valves. Permanently installed freon test probes with external connections are provided for testing the adsorber. A radiation monitor is located immediately downstream of the atmospheric cleanup equipment bypass.

2.2.2   BWR Only--The air removal equipment consists of two half-capacity mechanical vacuum pumps for initial evacuation of the main condenser and two full-capacity steam jet air ejectors (SJAE) for maintaining vacuum during normal operation.

Each vacuum pump is equipped with normally closed maintenance isolation valves and a discharge check valve. The pumps discharge into a common header which discharges to the turbine building ventilation system. A radiation monitor and alarm is supplied to detect high radioactivity of pump discharge.

Each steam jet air ejector is equipped with maintenance isolation valves and a suction side relief valve. The ejector discharges to the reactor off-gas system.

3.0 SYSTEM DESIGN REQUIREMENTS

3.1 CONDENSER SYSTEM

General requirements for performance under certain plant operating conditions are given below.

3.1.1 Startup (through turbine synchronization)

The condenser system shall heat and deaerate the initial inventory of water contained in the secondary cycle to meet startup temperature and oxygen content limitations.

The system shall accept recirculation flows, bypass flows, and miscellaneous drain and vent flows associated with secondary cycle evacuation, deaeration, cleanup, and heatup.

The system shall condense steam from the main and feed pump turbines.

Prior to initial startup, the main condenser may be used to store large quantities of chemical cleaning solutions and flushing water for construction cleaning and flushing of various plant equipment and piping.

3.1.2 Normal Operation (from initial turbine loading to maximum anticipated power level)

The system shall condense steam from the main and feed pump turbines and accept miscellaneous flows from other systems while providing the backpressures required.

The main condenser, in conjunction with the condenser vacuum system, shall remove non-condensible gases and deaerate the condensate to maintain the dissolved oxygen requirements necessary for unit operation.

3.1.3 Transient Conditions

The system shall provide the water inventory required by the feedwater and condensate systems in meeting their design transient conditions.

The system shall be designed to accept the main turbine exhaust, feed pump turbine exhaust, turbine bypass flow, and other miscellaneous flows for specified operating transients (e.g., step turbine load reductions).

BWR Only--The system shall be capable of accepting the moisture separator safety valve discharge flow.

162

3.1.4   Hot Standby

The system shall condense the feed pump turbine exhaust to maintain backpressures required.

The system shall also accept miscellaneous drain and vent, recirculation and bypass flows required for standby operation of the secondary cycle.

The system shall maintain required oxygen levels for the secondary cycle.

3.1.5   Shutdown

The system shall maintain condensate oxygen levels through deaeration until condenser vacuum is broken.

The system shall accept and condense steam from the feed pump turbines and turbine bypass system to maintain required operating backpressures until vacuum is broken.

The system shall accept miscellaneous drain and vent recirculation and bypass flows until their function is terminated.

The main condenser shall be used to store water during refueling operations.

3.2   CONDENSER VACUUM SYSTEM--PLANT OPERATING CONDITIONS

3.2.1   Startup

The system mechanical vacuum pumps shall be used for initial evacuation of the main condenser and feed pump turbine condensers. Pump discharge is unfiltered, but shall be continuously monitored for radiation contamination.

3.2.2   Normal Operation

PWR Only--During normal operation with up to the maximum expected air inleakage, the condenser vacuum system shall remove all noncondensibles and air inleakage necessary to provide proper condenser venting. Vacuum pump discharge shall be processed by the atmospheric cleanup equipment and monitored for radiation contamination.

BWR Only--During normal operation the system shall remove all noncondensible gases required to maintain condenser vacuum and shall deliver the off-gases at sufficiently high pressure

163

to be processed by the off-gas system and with adequate dilution steam to maintain the hydrogen content of the off-gas effluent to minimize potential for off-gas detonation in accordance with the reactor supplier's specifications.

3.2.3 Adverse Conditions

The most severe effect any adverse condition can have on the system is complete loss of system capability to provide proper condenser venting. This can be accommodated by a shutdown of the unit with the unit capable of being returned to operation after corrective action.

3.3 NUCLEAR HAZARDS REQUIREMENTS

The condenser and auxiliaries are not required to affect or support the safe shutdown of the reactor or perform in the operation of reactor safety features. Therefore, there is no need for provisions to preclude the loss of the system during abnormal operating conditions.

BWR Only--All condenser vacuum system failures, including any that may occur during or following the occurrence of any of the design basis natural phenomena for the safety related structures of the plant, shall be analyzed to determine if protective measures are necessary to preclude any danger to control room personnel or safety related equipment.

4.0 SYSTEM PERFORMANCE

4.1 MAIN CONDENSER PERFORMANCE REQUIREMENTS

Important main condenser performance parameters (e.g., back pressure and condensate outlet temperature depression) shall be guaranteed at the annual average cooling water temperature for cleanliness factors of 85, 95, and 100 percent. The duty of the main condenser for this purpose shall be based on the main turbine exhaust and feed pump turbine condensate drain flows and enthalpies given by the turbine-generator guarantee full-load heat balance. Energy from other streams shall be considered to be negligible.

The maximum cooling water pressure drop with commercially clean tubes measured between waterbox inlet and outlet shall be guaranteed by the manufacturer.

The condenser shall be guaranteed to deaerate the condensate to a maximum dissolved oxygen content of 7 ppb (under all normal operating conditions) for the turbine load range specified and for the range of cooling water temperature expected.

The air-vapor mixture temperature leaving the air cooler zone shall be at least 7.5°F below the saturation temperature corresponding to the pressure at the steam inlet during normal operation.

4.1.1   Dumps to Main Condenser

In case of malfunctions in the heater drains and vents system during startup, low loads, or upon sudden load reductions, various drains and vents will be dumped or routed to the main condenser. The condenser shall be designed to receive these abnormal streams without overheating or without excessive erosion or vibration of the condenser internals. (Their energies are not to be considered in condenser sizing or determination of performance.)

A turbine bypass system is supplied to provide flexibility during unit startup and power operation so that turbine load changes in excess of the NSSS design can occur without reactor scram (or lifting main steam safety valves). During turbine bypass, the condenser system shall be capable of accepting (1) the maximum bypass system capacity, and (2) the corresponding maximum available turbine exhaust flow without exceeding the maximum allowable differential pressure[1] between zones throughout the range of cooling water temperatures. The division of turbine bypass steam flow between the zones is determined accordingly. The maximum allowable pressure differential should not be exceeded with one cooling circuit (waterbox) isolated (should be considered only on an individual project basis).

The turbine bypass steam shall be routed and distributed to avoid overheating and excessive erosion or vibration of the condenser internals or the turbine itself. Interlocks shall be provided to prevent opening of the turbine bypass valves when there is danger of exceeding permissible turbine back-pressure or temperature limitations.

In a BWR system, accommodations shall be made for receiving the moisture separator/reheater relief valve discharge. This flow and the normal turbine exhaust flow will not occur simultaneously.

---

1. As determined by the more restrictive of the requirements of the turbine manufacturer or the condenser manufacturer.

4.1.2   Special Operating Conditions

During unit startup, the water inventory in the hotwell, condensate storage system, and condensate feedwater system will be circulated through the main condenser to obtain deaeration (maximum of 42 ppb of dissolved oxygen). A hotwell sparging system, using auxiliary steam or steam from the other unit, shall be designed to accomplish this initial deaeration within 8 hours after vacuum and a recirculation flow through the hotwell (6000-8000 GPM) are established. In addition, water for cleaning and flushing various plant piping systems will be heated to 180°F to 200°F in the condenser hotwell with sparging steam. Condenser material selection and design shall permit the initial fill of condensate to be as cold as 40°F.

The main condenser must be designed for waterbox isolation and the resulting increase in circulating water flow through the unisolated tube bundle.

The main condenser must be capable of accepting increased turbine exhaust flow at stretch power conditions without excessive erosion or vibration.

4.2   FEED PUMP TURBINE CONDENSER PERFORMANCE REQUIREMENTS

The operating backpressure shall be guaranteed at the annual average cooling water temperature for a cleanliness factor of 85 percent. The condenser duty for this purpose is to be based on the turbine operating at 100 percent full load (normal operating conditions) at the guaranteed backpressure.

The maximum cooling water pressure drop including nozzles with commercially clean tubes shall be guaranteed by the manufacturer.

In addition the operating backpressure at the hottest day cooling water temperature and with two feedpumps operating at 100 percent full load shall be verified to be within turbine allowable exhaust pressure limitations for a cleanliness factor of 85 percent.

4.3   CONDENSER VACUUM SYSTEM PERFORMANCE REQUIREMENTS

4.3.1   PWR Only

All three mechanical vacuum pumps will operate during startup, and shall be capable of evacuating the main condenser (including equipment vented to the main condenser) down to condenser operating pressures. Each pump shall have a minimum guaranteed capacity of 800 scfm at a suction pressure of 15 in.HgA for

166

this purpose (hogging operation). Two vacuum pumps will normally operate during unit power operation to remove the expected air inleakage of 15 scfm at 1 in.HgA suction pressure (holding operation). Three vacuum pumps will operate during unit power operation with abnormally high air inleakage, and shall be capable of removing the design air inleakage and noncondensibles of 45 scfm with a rated suction pressure of 1 in.HgA.

The vacuum pump exhaust processing equipment (heater and absorber) shall be capable of processing the system holding capacity at the maximum expected condenser LP zone pressure (see section 5.1) during normal operation.

System effluent shall be continuously monitored during all modes of operation to detect gaseous radioactivity levels.

The condenser vacuum system shall be capable of handling the feed pump turbine condenser vent flows. These vents will be piped to both the suction of the vacuum pumps and the main condenser LP zone. During normal operation, normally closed isolation valves shut off this direct connection between the air removal equipment suction header and the feed pump turbine condenser vents. Feed pump turbine condenser venting is through the main condenser.

4.3.2   BWR Only

Each steam jet air ejector (SJAE) shall be capable of removing all noncondensible gases required to maintain normal operating vacuum (holding operation). Each SJAE shall have a minimum guaranteed capacity of 4S SCFM dry air, plus additional non-condensibles as determined by the NSSS supplier, plus steam to saturate the total noncondensible mixture. Only one SJAE for each unit shall be in operation at a given time. To minimize the potential for a detonation in the off-gas system upstream of the catalytic recombiner, sufficient dilution steam shall be maintained in the SJAE discharge to limit the hydrogen concentration to less than four percent by volume.

The mechanical vacuum pumps shall be capable of initially evacuating the condenser system (including equipment vented to the main condenser) from atmospheric pressure to SJAE operating pressures. Both pumps of a unit shall be required to operate in parallel. Each shall have a minimum guaranteed capacity of approximately 1100 scfm at 15 in.HgA for initial evacuation.

The mechanical vacuum pumps shall not be permitted to operate when the reactor pressure exceeds a specified value in order to limit radioactive release to the atmosphere. (The pumps discharge to the atmosphere via the turbine building ventilation system.)

The condenser vacuum system shall be capable of handling the feed pump turbine condenser vent flows. These vents will be piped to both the suction of the air removal equipment and the main condenser LP zone. During normal operation, normally closed isolation valves shut off this direct connection between the air removal equipment suction header and the feed pump turbine condenser vents, thus feed pump turbine condenser venting is through the main condenser.

5.0 SYSTEM COMPONENTS

5.1 MAIN CONDENSER

The condensers shall be horizontal, single shell, divided waterbox, longitudinal, single pass, triple pressure, surface type. Each condenser shall have its total surface equally divided between the three pressure zones, designated "low pressure (LP) zone," "intermediate pressure (IP) zone," and "high pressure (HP) zone." There shall be a middle waterbox (vertically partitioned into two equal sections) to reduce the required tube lengths. There are three turbine exhaust connections, one for each double flow, low pressure turbine. There will be three condenser necks corresponding to the three turbine exhausts, one for each pressure zone. Condensate will flow from the LP zone hotwell to the IP zone hotwell through a loop seal and in the same manner from the IP zone hotwell to the HP zone hotwell. The HP zone hotwell will extend essentially the full length of the condenser beneath the LP and IP zones. The three hotwell pumps take suction from the HP zone hotwell.

BWR Only--The design shall include a holdup hotwell to accomplish the required radionuclide decay. Condensate will flow from the LP zone hotwell to the IP zone hotwell through a loop seal and in the same manner from the IP zone hotwell to the HP zone hotwell. The condensate will then flow from the HP zone hotwell to a holdup hotwell located beneath the zone hotwells and extending essentially the full length of the condenser. The three hotwell pumps take suction from the holdup hotwell.

Three low-pressure feedwater heaters will be mounted in each condenser neck.

Each condenser shall have a total condensing surface based upon optimization of the heat rejection system. (This optimization, however, must be consistent with the fixed condenser parameters defined herein.) Condenser configuration shall be consistent with any size and space limitations imposed by plant layout.

5.1.1 Condenser Shell and Expansion Provision

The design pressure of the shell shall be 30" Hg vacuum and suitable for an emergency internal pressure of 20 psig with an allowance for static head developed during hydrotest. Shell material shall be steel plate. The shell shall be well braced against injurious strains and distortions. It shall be designed for a minimum number of field joints consistent with transportation facilities.

The condenser shall be anchored near the middle so that it expands toward each end and shall have lubrite base plates or other accommodations for expansion to reduce the resulting forces on the condenser supports to a minimum under the most adverse operation conditions.

The design shall allow for differential expansion between the tubes and the shell based on initial temperatures of 50°F and maximum temperature (for both shell and tubes) of 200°F.

The condenser shall have supports, designed to rest on the concrete foundation and support the fully flooded weight of the condenser without the addition of temporary shoring or supports.

5.1.2 Condenser Tubes

The condenser tubes will be 90-10 Cu-Ni. The tube gauge, diameter, and length shall be consistent with the optimization of the heat rejection system. The tubes shall be rigidly expanded into the tube sheets at both ends and shall be belled or flared at the inlet ends to provide a smooth flow of water as it enters the tubes. The tubes shall be arranged for self-draining by sloping from the middle toward each end. The centerline of each tube shall be a straight line. Water velocity in the tubes shall be 6 to 8 feet per second.

5.1.3 Condenser Tube Sheets and Tube Joints

The tube sheets shall be of free-machining steel plate. The tube sheets may be welded to the shell. The tube holes in the tube sheet shall be grooved.

169

5.1.4   Tube Support Plates

Sufficient steel support plates shall be furnished so that
there will be no sag in the tubes. These plates shall be
welded to the shell and shall be located to prevent vibration.
Spacing between tube support plates shall not exceed 36 inches.

5.1.5   Air Cooler Section

The air cooler section shall be sized so that the air-vapor
mixture temperature leaving the air cooler zone shall be
at least 7.5°F below the saturation temperature corresponding
to the absolute pressure at the gas-vapor offtake connection
when the condenser is operating such that this pressure is
1.05 inches mercury absolute (in.HgA). A separate air take-
off connection shall be provided for each tube bundle to
facilitate location of air leaks.

5.1.6   Pressure Zone Partition

Steel pressure zone partition plates shall be provided to
maintain the differential pressure between zones. The plates
shall be welded to the wall of the condenser shell and shall
be designed to withstand the maximum pressure difference
between adjacent condenser zones during all modes of condenser
operation (including steam dump).

5.1.7   Waterboxes and Covers

The waterboxes shall be designed for a pressure compatible
with the condenser circulating water system design pressure,
and need not be designed for vacuum. A vacuum breaker will
be installed to prevent accidental subjection of the waterboxes to vacuum. The waterboxes shall be designed so that
the water velocity in each tube is approximately the same.
All intermediate and outlet waterbox compartments shall be
provided with a 4-inch flanged connection on top of the box
for air release.

Approved tubular gauge glasses, complete with shutoff cocks,
shall be furnished on each inlet and outlet waterbox. A
tapped opening for attaching a float switch for operating
the air release valve shall be furnished on the top section
of each intermediate and outlet waterbox. The waterboxes
shall be designed to be self-supporting from the main condenser shell. Due to the possible installation of a condenser
tube cleaning system, the inlet waterbox and the outlet waterbox shall have ample strength to support its weight in addition
to a portion of the weight of the condenser circulating water
conduit. Also, ample space shall be provided around inlet
and outlet conduits for the tube cleaning system.

All inlet and outlet waterboxes shall be supplied with
sufficient manhole openings (at least one 20-inch and one
24-inch opening in each compartment) with quick-opening
hinged lids. The securing mechanism for lids shall be
capable of being operated by hand or by striking with a
hammer (e.g., lever nuts, dogs). Each of the inlet and
outlet waterboxes shall be supplied with an inside ladder
permanently mounted and accessible through the 24-inch
manhole opening.

The intermediate waterbox shall be vertically for isolation
of water circuits. Each compartment shall have a dewatering
drain and shall be provided access from the side through two
36-inch diameter manholes with quick-opening, hinged lids.

All waterboxes shall be provided with one-inch tapped and
plugged connections for use in connecting a test U-tube.

Both inlet and outlet waterboxes shall be provided with bolted
covers for retubing.

Each waterbox connection 18-inches diameter and larger that is
oriented vertically downward shall have a grating for personnel
protection.

All connections that may trap condenser tube cleaning balls
shall be provided with a screen of not less than 3/8-inch
mesh to prevent such entrapment.

5.1.8   Hotwells

All hotwells shall be of welded steel plate and constructed
to withstand full vacuum and hydrostatic pressure (without
cribbing) when the condenser is flooded up to the turbine-
condenser connection. Each hotwell shall be fitted with
full-length tubular gauge glasses, complete with shutoff
cocks with ball checks for vacuum sealing, root valves
suitable for attachment of tygon tubing for hotwell pump
level measurement, and with one 24-inch manhole with a
quick-opening, gasketed, hinged lid.

For early detection of tube leaks, the zone hotwells shall
be divided and sample connections shall be provided to seg-
regate condensate produced by each tube bundle and bring
water out of each zone to the sampling system.

The condenser manufacturer shall set the depths of the zone
hotwell sections to properly drain by gravity from the LP zone
to the IP zone and from the IP zone to the HP zone under the
most severe differential pressure conditions and to achieve
the maximum possible reheating.

A 4-inch high "dam" or other suitable device (other than a screen) shall be located at each outlet to prevent nonfloating debris from entering a hotwell pump. Also, a cover-type baffle shall be installed at each outlet connection to prevent objects from dropping into the outlet.

The condenser hotwells (or shells) shall have baffles to prevent incoming drains from upsetting the level controls (triggering false readings).

The hotwell shall be equipped with a steam sparging system for deaerating condensate during unit startup. The headers and distribution piping shall be carbon steel; sparger nozzles shall be stainless steel.

PWR Only--The HP zone hotwell shall have a storage capacity equivalent to 2-1/2 minutes of residence time at full load operation above the "dam."

BWR Only--The condenser manufacturer shall design the holdup hotwell to provide at least four minutes of residence time at full load operation for radionuclide decay of the condensate. The holdup hotwell shall have this storage capacity above the "dam."

5.1.9   Connections

In addition to the principal connections, ancillary connections shall be installed on the condenser to receive drains, vents, recirculation flows, makeup, and other inputs. These inputs originate from various component equipment of the nuclear plant, such as feedwater heaters, moisture-separator-reheaters, steam piping, pumps, startup system dumps, miscellaneous heat exchangers, gland leakoffs, and condensate from pump drive turbines. The number, size, location, and arrangement of all connections are determined by the turbine plant design. These connections will be developed after the condenser is purchased. Thermal sleeves shall be used at condenser wall penetrations for all high temperature drains and vents (MFP recirculation line, heater drain bypass lines, heater shell relief vents, etc.).

All termination points for incoming flows to the condenser must be above the maximum water level in each zone.

Spray headers and baffles, sized to minimize upstream flashing and prevent tube erosion, are required. The condenser manufacturer shall submit detailed drawings of the internal spray header and baffle arrangement for review; a pre-operational visual inspection will be undertaken to verify the adequacy of the design.

Continuous drains with usable energy should preferably be routed to the HP zone.

Drains requiring deaeration shall be introduced above the centerline of the tube bundle, if possible.

5.1.10 Condenser Necks and Expansion Joints

Separate condenser necks are required for each pressure zone. The condenser necks shall be steel plate constructed for full vacuum and hydrostatic pressure when condenser is flooded up to the turbine-condenser connection, and well braced against injurious strains and distortions. The necks shall be designed for a minimum number of field joints consistent with shipping limitations.

Each condenser neck pressure zone section located below an expansion joint shall have a vacuum-tight bolted manhole with minimum clear opening of three by three feet, and a 20-inch minimum quick-opening manhole. The joint between the condenser neck and turbine exhaust hood shall be designed for field welding. A landing bar shall be provided on top of the neck for making the connection with the turbine. Condenser expansion joint loads on turbine exhaust nozzle shall be consistent with allowable turbine exhaust nozzle loading.

Expansion joints of the rubber-belt type (a continuous band of reinforced rubber) shall form the perimeter of the connection between the upper and lower portions of each condenser neck. The joints are to be connected by means of bolted clamps to end pieces which will be welded to the condenser neck. The joints shall be suitably backed to prevent direct impingement of incoming steam and arranged to offer minimum resistance to steam flow. They shall be designed to prevent the leakage of air into the condenser. A water seal arrangement for the rubber expansion joint shall be supplied. A stainless steel expansion joint may be offered as an alternate, in which case a water seal will not be required. Water for the expansion joint water seal shall be taken from the gland seal water system.

Adequate supports and provision for expansion shall be furnished for three LP heaters in each of the condenser necks. Supports for extraction piping may be required.

The expansion joint, struts, piping, and structural members in the condenser neck shall be designed and located to minimize disturbance to the turbine exhaust flow.

173

5.1.11 Air Removal

PWR Only--The condenser shall be designed for efficient removal of noncondensable gases to produce optimum condenser pressures and specified dissolved oxygen levels. Three mechanical vacuum pumps per condenser will be supplied to remove noncondensibles. During normal operation, condenser venting will be accomplished by the condenser vacuum system.

BWR Only--The condenser shall be designed for efficient removal of noncondensible gases, including those resulting from radiolytic dissociation, to produce optimum condenser pressures and specified dissolved oxygen levels. During normal operation, condenser venting will be accomplished by the condenser vacuum system.

5.1.12 Manifolds

Grouping or manifolding of streams to the condenser shall be minimized. Only lines of similar function and operating pressure are to be manifolded.

The cross-sectional area and layout of each manifold shall have a manifold pressure that is low enough (with all streams coming in) to avoid limiting or restricting the flow of any incoming stream (e.g., critical pressure drops across orifices and certain flow control devices should always be available). The lines coming into the manifolds should be arranged in descending order of pressure, with the highest pressure line coming in farthest from the manifold opening to the condenser.

Streams from different sections of a line or pieces of equipment should not be manifolded if they experience extreme differences in pressure during transient or abnormal operation.

Continuous water drains shall not be routed to manifolds.

Pipes discharging steam from turbine bypass valves to the condenser are not to be manifolded.

5.1.13 Instrumentation

Instrumentation and controls shall be provided for the following logic, control, and indication functions:

a. Hotwell and waterbox level indication

b. Automatic maintenance of hotwell level

c. High back pressure alarm and main turbine trip

d. Vacuum breaker actuation

e. Interlocking of turbine bypass valves

f. Tube pressure differential as required for condenser circulating water system operation

g. Back pressure signals for starting and stopping of air removal equipment.

5.1.14 Test, Maintenance, and Inspection Criteria

5.1.14.1 Testing Criteria

The shell shall be hydrostatically tested in the field. An ASME performance test may be performed on a selected basis.

5.1.14.2 Maintenance and Inspection Criteria

Periodic functional tests and visual inspections of the main condenser will be performed. Inservice inspection (ASME Code, Section IX) is not required.

The design shall include provisions for waterbox isolation, tube removal and plugging, and personnel access to hotwells and shell.

Drain connections shall be installed to permit removal of water from internals to the building sump.

5.2 FEED PUMP TURBINE CONDENSER AND VENT PIPING

Each feed pump turbine condenser shall be a horizontal, surface type. It shall operate without a water level. Condensate shall be drained by gravity through a loop seal to the main condneser. (The elevation of feed pump turbine condenser must permit gravity drains.) The condenser base will be rigidly supported, with an expansion joint between the condenser and the turbine exhaust connection.

Each condenser shall have a total condensing surface based on 2" HgA backpressure. See Design Guide DG-M2.19, "TURBINE GENERATOR HEAT CYCLE--Main Feed Pump Turbines" for calculations. The condenser configuration shall be consistent with any size and space limitations imposed by plant layout.

5.2.1 Shell and Neck

The design pressure shall be 20 psig and 30" Hg vacuum. The necks shall have two 24-inch manholes with quick-opening, gasketed, hinged covers. A landing bar shall be located at the top of the neck for making the connection to the turbine. An expansion joint shall be installed in the condenser neck.

A circumferential stainless steel plate for tube protection shall be located at any point where the shell must be cut for tube removal. The number, sizes, and locations of all shellside connections will be calculated after purchase of the turbine and condenser. A maximum temperature differential of 200°F between tubes and shell shall be used for differential expansion design.

Each condenser shell shall be individually vented to the low pressure zone of the main condenser. The vent line from each feed pump turbine condenser requires a maintenance isolation valve. At a point between this isolation valve and the vent connection on the feed pump turbine condenser shell, the vent line shall be tied into the suction header of the air removal equipment for leak testing each feed pump turbine condenser shell.

5.2.2   Condenser Tubes

The tubes will be made of 90-10 Cu-Ni, rigidly expanded into the tube sheets at both ends and belled or flared at the inlet ends. The tubes shall be arranged for self-draining by sloping toward outlet end.

5.2.3   Condenser Tube Sheets and Tube Joints

Tube sheets shall be free-machining steel plate. They may be welded to the shell. The tube holes in the tube sheets shall be grooved.

5.2.4   Tube Support Plates

All condensers shall have sufficient support plates located at intervals which will prevent tubes from vibrating or sagging. Tube holes shall be accurately spaced and fabricated to prevent cutting or chafing of the tubes. Tube support plates shall be welded to the shell.

5.2.5   Waterboxes and Covers

The waterboxes shall be designed for a pressure compatible with the condenser circulating water system design pressure and need not be designed for vacuum. Vacuum breakers will be installed to prevent accidental subjection of the waterboxes to vacuum.

The waterboxes and waterbox covers shall be fabricated steel or cast iron, and shall be self-supporting from the condenser shell. The top of each waterbox shall have a 2-inch flanged connection for air release. Approved tubular gauge glasses, complete with shutoff cocks, shall also be furnished. Gauge glasses shall be located so that water level can be determined in all passes and shall cover the entire waterbox height. A tapped opening for attaching a float switch to operate the air release valve shall be located in the top section of each waterbox.

The waterboxes shall be supplied with sufficient manhole openings with quick-opening hinged covers to permit inspection or plugging of both ends of all tubes. The securing mechanism for covers shall be capable of being operated by hand or by striking with a hammer (e.g., lever nuts, dogs). Access openings in the pass partition plates are not acceptable.

Waterbox design shall permit access for complete retubing.

Waterboxes shall have sufficient 1-inch tapped and plugged openings for connecting test U-tubes for measuring the pressure drop across all tube passes.

Each waterbox shall have a two-inch dewatering drain connection, prepared for socket welding. A sentinel type water relief valve will also be installed on each waterbox.

5.2.6   Expansion Joints

Expansion joints shall be stainless steel. Each expansion joint shall form the perimeter of the connection between the turbine exhaust and the condenser neck. The joint between the condenser neck and the turbine exhaust hood shall be designed for field welding. Landing bars shall be located on top of the necks for making the connections with the turbine. Expansion joints shall be suitably backed to prevent direct impingement of incoming steam and arranged to offer minimum resistance to steam flow. Peripheral welds of the expansion joints shall not be permitted.

The joints shall be capable of absorbing up to 3/8-inch of axial movement and 3/8-inch lateral movement. Lateral bracing of the expansion joints shall be the responsibility of the contractor.

5.2.7   Instrumentation

Instrumentation and controls shall be provided for the turbine back pressure alarm and trip function and the low level alarm function.

5.2.8   Test, Maintenance, and Inspection Criteria

5.2.8.1   Testing Criteria

The condenser will undergo shop tests as required by the specification. There are no specific field test requirements.

5.2.8.2   Maintenance and Inspection Criteria

Periodic functional testing and visual inspections will be
carried out. Inservice inspection as defined by the ASME
Code, Section XI is not required. The design shall include
provisions for cooling water and steam side isolation, tube
removal and plugging, and personnel access to the neck.

5.3   CONDENSER VACUUM SYSTEM COMPONENTS

5.3.1   Mechanical Vacuum Pumps

The vacuum pumps shall be liquid-ring design. The compressant (liquid used to form the compressor "ring") shall
be supplied from the gland seal water system and shall be
recirculated through a heat exchanger (supplied with vacuum
pump) in order to minimize compressant makeup. Cooling water
to the heat exchanger shall be supplied by the raw cooling
water system.

Each pump shall be supplied with an inlet valve (fail closed)
to automatically isolate the pump from the system when not
in service.

The vacuum pump motors shall operated on 3 phase, 60Hz, 480V
station power.

Cooling water for the SJAE intercondensers is supplied by
main condensate. The intercondenser tubes shall be 304SS,
welded to the tubesheets; the intercondenser channel side
design pressure shall be compatible with that of the appropriate section of the condensate system. Tube sheets shall
be SS or SS-clad carbon steel. The shell shall be welded
steel and designed for full vacuum and a maximum pressure,
assuming hydrogen detonation. Steam condensed by the intercondenser will be returned to the main condenser.

5.3.2   Piping and Valves

The piping (or ducts) and valves shall be sized to pass the
required flows and be designed for the following minimum
pressures and temperatures:

Air suction - (PWR) 50 psig/30 in. Hg vacuum, 200°F
            - (BWR) base on hydrogen detonation

Air discharge - (PWR) 50 psig/175°F
              - (BWR) base on hydrogen detonation

The feed pump turbine condenser vent piping and valving shall be designed for 50 psig/full vacuum, 200°F or for hydrogen detonation in the case of a BWR, whichever is higher.

ANSI B31.1 shall control the fabrication and erection of piping and valves in the condenser piping system. Low point drains shall be provided.

For BWR's only, all valves shall have spark free valve seats.

5.3.3   Atmosphere Cleanup Equipment (PWR Only)

All air cleanup equipment and related instrumentation for PWR systems shall be designed, tested, and maintained in accordance with applicable NRC guidelines except where specific exceptions are taken.

5.3.3.1   Air Heater

A heater shall heat the vacuum pump exhaust to reduce its relative humidity from 100 percent to 70 percent or less for all operating temperatures between 40° and 115°F prior to reaching the adsorbers. Controls are not required to maintain this value.

5.3.3.2   Adsorption Units

A carbon adsorber unit shall be installed to remove gaseous iodine from the air stream. The adsorbent beds shall be a minimum of two inches deep.

5.3.4   Power Supply

Power shall be supplied from the normal ac auxiliary power system.

5.3.5   Instrumentation and Controls

In addition to the instrumentation required to perform the logic functions, the following instrumentation and centrals shall monitor and control the system operation:

a. Local indication

    1. Mechanical vacuum pump test discharge flow supplied with pump.

    2. Pump suction pressure test connection.

3. Pump discharge temperature.

4. Pump compressant temperature in and out of heat exchanger (supplied with pump).

5. Pump discharged pressure.

6. (PWR only) Pressure differential across charcoal adsorber.

7. (PWR only) Temperature out of preheater.

8. (BWR only) SJAE discharge pressure.

9. (BWR only) Offgas flow.

10. (BWR only) Pressure at inlet of SJAE test connection.

11. Radiation level of vacuum pump discharge flow.

b. Main control room (MCR) display

1. Mechanical vacuum pump status (on-off).

2. (PWR only) Flow and temperature upstream of radiation monitor.

3. (BWR only) SJAE discharge pressure.

4. Valve position for all remote operated valves.

5. (BWR only) Radiation level of vacuum pump discharge (to be recorded).

c. MCR alarm

1. High radiation level (see item a.10).

2. (PWR only) Vacuum pump trip.

3. (PWR only) High adsorber differential pressure (see item a.6).

4. (PWR only) Low temperature after preheater (see item a.7).

5. (BWR only) High reactor pressure while mechanical vacuum pump is operating.

d. Automation

  1. Automatic operation of the vacuum pump suction
      valve (provided with the pump).

  2. Local and remote (in the MCR) manual start and stop
      of each pump.

  3. (PWR only) Local and remote (in the MCR) manual
      opening and closing of the air cleanup equipment
      bypass valve.

  4. (BWR only) Local and remote (MCR) manual control of
      the SJAE discharge valve.

  5. (BWR only) Trip and isolate mechanical vacuum pump
      when the radiation level in the effluent is high or
      reactor pressure is high.

5.3.6 Logic for Operation of Condenser Vacuum System

5.3.6.1 PWR Only

Initially, the air cleanup bypass valve is opened and all
three vacuum pumps are placed in operation. Any one of
the three pumps can and is to be selected for automatic
start-stop operation. This pump will stop on low condenser
HP zone pressure; it will automatically restart on high
HP zone pressure and remain running until manually shut
off. Start and stop controls for each pump should be
located locally and in the MCR.

Two pumps are operated during normal operation (HP zone
pressure below high pressure setpoint) of the system, with
the third pump in standby. If the pressure increases above
the high pressure setpoint, the standby pump automatically
starts. Once started, this pump can only be manually stopped.
The air cleanup equipment bypass valve is open, with system
effluent being bypassed around the carbon adsorber.

Upon determination that condenser offgas radiation levels
exceed normal limits, the bypass is manually closed (initiated
in the MCR) to divert the flow through the preheater and
charcoal adsorber. The preheater is warmed up prior to
admitting air to the cleanup equipment.

5.3.6.2 BWR Only

Two mechanical vacuum pumps will be started initially for
hogging operation. These pumps will be stopped when the

condenser HP zone pressure is low enough to allow operation of the SJAE. The pumps are not allowed to run if the reactor pressure is high enough that the condenser dump valves will open in the event of a scram.

One of two SJAE's is sufficient for normal operation and only one SJAE per unit shall be in operation at a given time. On rising backpressure, the operating SJAE will trip (isolate) and the standby SJAE is automatically placed in operation.

The isolation-start sequence is:

a. Close steam supply isolation valves to running SJAE.

b. Close isolation valve at SJAE suction.

c. Open steam supply isolation valves to standby SJAE.

d. Open isolation valve at SJAE suction after the SJAE steam supply isolation valve is fully open.

Run permissives for the SJAE are (1) condensate flow through the intercondenser, (2) water level less than high and (3) main/auxiliary steam available.

5.3.7   Test, Maintenance, and Inspection Criteria

5.3.7.1   Testing

Each heat exchanger, valve, SJAE and vacuum pump shall receive a shop hydrostatic test which shall be performed according to applicable codes.

In addition, a complete noncritical preoperational test shall be performed to verify the proper functioning of the components of the system.

PWR Only--Testing requirements of atmospheric cleanup equipment are covered in latest Branch Technical Position - ETSB, "Design, Testing, and Maintenance Criteria for Normal Ventilation Exhaust System Adsorption Units of Light Water-Cooled Nuclear Power Reactor Plants."

5.3.7.2   Maintenance and Inspection

Since no component has a code safety class, inservice inspection as defined in ASME Boiler and Pressure Vessel Code, Section XI, is not required. However, periodic functional tests will be performed in conjunction with unit outages.

The system design shall permit access to all components
for visual inspection and on-line maintenance. Access ways
and lifting devices shall be provided to permit removal
and replacement of equipment. All equipment shall have
provisions to permit complete removal of water from internals.

PWR Only--The Branch Technical Position (referenced in
section 5.3.7.1) covers maintenance requirements applicable
to atmospheric cleanup equipment.

BWR Only--Facilities in the turbine building shall permit
decontamination of equipment that must be removed from the plant.

6.0 SYSTEM INTERFACES

The following documents will assist in defining system interfaces
and provide additional insight into the design of a condenser system
for a 3800 MWT nuclear plant with light water reactors:

a. EN DES Design Guide DG-M2.3.1, "TURBINE GENERATOR HEAT CYCLE--Condensate
    System - Nuclear," TVA; Knoxville, TN: 1979.

b. EN DES Design Guide DG-M2.1.1, "TURBINE GENERATOR HEAT CYCLE--Main
    Steam System - Nuclear," TVA; Knoxville, TN: 1979.

c. EN DES Design Guide DG-M2.6.1, "TURBINE GENERATOR HEAT CYCLE--Condensate
    Storage and Transfer System - Nuclear," TVA; Knoxville, TN: 1979.

d. EN DES Design Guide DG-M3.1, "DESIGN CONCEPTS OF LOW PRESSURE
    SYSTEMS--Heat Rejection System," TVA; Knoxville, TN: 1979.

e. EN DES Design Guide DG-M2.2.1, "TURBINE GENERATOR HEAT CYCLE--Main
    Feedwater System - Nuclear," TVA; Knoxville, TN: 1979.

f. EN DES Design Guide DG-M2.5.1, "TURBINE GENERATOR HEAT CYCLE--Heater
    Drains and Vents - Nuclear," TVA; Knoxville, TN: 1979.

g. EN DES Design Guide DG-M2.9, "TURBINE GENERATOR HEAT CYCLE--Condensate
    Polishing Demineralizers," TVA; Knoxville, TN: 1979.

h. EN DES Design Guide DG-M2.8.1, "TURBINE GENERATOR HEAT CYCLE--Auxiliary
    Steam System - Nuclear," TVA; Knoxville, TN: 1979.

i. EN DES Design Guide DG-M2.10.1, "TURBINE GENERATOR HEAT CYCLE--Control,
    Service, and Low-Pressure Breathing Air Systems - Nuclear," TVA;
    Knoxville, TN: 1979.

j. EN DES Design Guide DG-M2.15, "TURBINE GENERATOR HEAT CYCLE--Chemical
    Treatment System/Equipment for Condensate-Feedwater and Raw Cooling
    Water Systems," TVA; Knoxville, TN: 1979.

Chapter 7
Design Guide DG-M3.1.1: Heat Rejection System

INTRODUCTION

This design guide represents current TVA design philosophy for the heat rejection system of a 3800 MW_t nuclear plant with a light water cooled reactor. It is intended to be an aid for the design of the heat rejection system and not a substitute for good engineering practice. An engineer with a thorough understanding of the system should prepare the design criteria and perform the design calculations.

The information is presented in two parts:

>   Part I: <u>Design Criteria</u>--enables a draft design criteria document
>   to be prepared for a specific project.

>   Part II: <u>Design Calculations</u>--sets forth the calculations required
>   for the basic design of the heat rejection system.

In preparing a design criteria for a specific project, this design guide should be used as the basis for the draft.

Part I: Design Criteria

1.0 GENERAL

1.1 SCOPE

This document, in conjunction with a specific TVA design criteria diagram and a specific TVA control logic diagram, is intended to be the governing document for design of the heat rejection (HR) system. This document provides general criteria and establishes design parameters for the system and for individual components of the system. The HR system may have both safety-related and non-safety-related requirements. However unless expressly stated, all requirements herein are non-safety-related. This document provides criteria to prevent damage to safety-related plant features as a result of a failure or malfunction of the system. If discrepancies exist between this document and any other system description or diagram, this document shall govern.

These requirements are based on two unitized HR systems for a two-unit power generating plant, except where noted. This design guide does not apply beyond the interface between the HR system and other systems and subsystems, except that this design guide shall apply to the makeup and blowdown systems.

1.2 ABBREVIATIONS

bhp	- Brake horsepower
Btu	- British thermal units
C	- Cycles of concentration in the system
CCW	- Condenser circulating water
CTM	- Cooling tower makeup
CWT	- Cold water temperature
E	- Energy, Btu/hr
ERCW	- Essential raw cooling water
°F	- Degrees, Fahrenheit
fcv	- Flow control valve
gal/min	- Gallons per minute
h	- Enthalpy, Btu/lb
hr	- Hour
HR	- Heat rejection
HRS	- Heat rejection system
H_2O	- Water
HWT	- Hot water temperature
I&C	- Instrumentation and controls
Intk PS	- Intake pumping station
lb/in^2g	- Pounds per square inch, gauge
m	- Mass flow, pounds mass per hour
MC	- Main condenser
mcr	- Main control room

```
mfpt    - Main feed pump turbine
npsh    - Net positive suction head
OEDC    - Office of Engineering Design and Construction
ppm     - Parts per million
Q       - Rate of heat transfer, Btu/hr
RCW     - Raw cooling water
RSW     - Raw service water
```

2.0 FUNCTIONAL DESCRIPTION

2.1 PURPOSE

The HR system shall be designed to provide an efficient means of dissipating waste heat from the power generation cycle primarily by evaporation to the atmosphere while meeting all applicable chemical and thermal criteria.

2.2 GENERAL DESCRIPTION

The plant consists of two generating units, each with a unitized (except for makeup and blowdown) HR system designed for continuous closed cycle operation. The main heat load results from condensation of steam in the main condenser (MC) which transfers the heat to the condenser circulating water (CCW). Secondary heat loads result from the raw cooling water (RCW) system, which supplies cooling water for various services and systems, and from the condensation of steam in the main feedpump turbine (mfpt) condensers, which are cooled by CCW in parallel with the main condenser. Table 1, "Heat Loads and Flows," is needed for the design criteria for each project. The blanks are to be filled in with information obtained from the design calculations (see Part II, "Design Calculations"), of this design guide.

TABLE 1
Heat Loads and Flows

COMPONENT/SYSTEM	FLOW (gal/min)	HEAT LOAD (Btu/hr)
Main Condenser		
Main Feedwater Pump Turbine Condenser		
Raw Cooling Water System		
TOTAL		

The system is made up of: cooling towers; a CCW pumping station; CCW pumps and associated bearing lubrication systems; cooling tower makeup (CTM) pumps and strainers located at the intake pumping station (Intk PS); blowdown pumps (if required); CCW conduits; main steam condenser and mfpt condenser water boxes, mechanical expansion joints; system valves; system makeup and blowdown conduits; desilting pond; blowdown holdup pond (if required); blowdown diffuser; and instrumentation and controls necessary to operate and monitor the system. A closely related component is the main condenser tube cleaning equipment, primarily a vendor-designed system.

Four CCW pumps are installed and are designed to operate in parallel[1] to circulate the CCW from the CCW pumping station through the supply conduits, condensers, and discharge conduit to the cooling tower where the heat is dissipated to the environment. The cooled water in the tower then flows to the CCW pumping station for recirculation through the system.

Gland seal water and bearing lube water for the CCW pumps are essential for continued operation. They are normally supplied by a tap off the CCW pump discharge. This water must be passed through a cyclone separation unit before injection into the CCW pumps. An alternate supply of gland seal water and bearing lube water should be available from the raw service water (RSW) system. Cooling water to the pump motors is normally supplied by a tap off the CCW pump discharge.

Evaporation, drift, and blowdown losses are replaced by the CTM system. Makeup is supplied from the river by CTM pumps after being filtered through traveling screens at the intake pumping station.

Enroute to the HR system injection point, part of the CTM flow may be diverted to the turbine building for use in the RCW system, with the remainder of the flow bypassing the turbine building. After use in the RCW system this water would be discharged into the makeup line and, in combination with that portion bypassed around the turbine building, will constitute the total volume of HR system makeup. In this event, a modulating valve would be required in the turbine building bypass line to assure that an adequate supply of water is diverted to the RCW system.

------

1. Normally, only three pumps are operated at one time.

Depending upon the design basis for the ultimate heat sink, the CTM pumps may provide a backup supply of cool water to the essential raw cooling water (ERCW) system after the loss of function of the primary ultimate heat sink as the result of a tornado. In this event, the normal makeup supply headers to the HR system would be isolated, diverting makeup flow to the ERCW system.

Water may also be supplied to the RCW system from the discharge of the CCW pumps upstream of the main condenser. In this event the flow would normally be reinjected into the CCW conduit downstream of the main condenser.

Each unit shall have one single-shell, multi-pressure, main condenser with a divided water box and provision for isolation of each water box for access and maintenance. Each unit shall also have three single-shell, single-pressure, mfpt condensers, with provisions for isolation of each condenser for access and maintenance.

The main condenser shall be equipped with an automatic condenser tube cleaning system which injects sponge-rubber balls into the CCW conduit upstream of the main condenser and collects the balls via a strainer in the CCW discharge piping.

3.0 GENERAL DESIGN REQUIREMENTS

3.1 ENVIRONMENTAL CONDITIONS

3.1.1 Floods

The system shall be operational up to and including administrative shutdown of the plant during the probable maximum flood.

3.1.2 Wind

Wind loads on the structures of this system, except the cooling towers, shall be in accordance with the design criteria document entitled "General Criteria for Design of Civil Structures."

3.1.3 Tornado

The system shall not be designed to withstand tornados. However, if the CTM pumps furnish a backup supply of cooling water for the ERCW system after the loss of function of the primary ultimate heat sink because of a tornado, limited tornado precautions are needed. In such a case, the intake pumping station shall be designed to protect the traveling screens, screen wash pumps, CTM pumps and strainers, and all associated piping and controls from the effects of a tornado. Precautions may be made to withstand the atmospheric pressure transients associated with the tornado either by specifying that the equipment essential for this mode of operation shall be designed to withstand those transients, or by designing the station to protect the equipment from the effects of the pressure transients.

3.1.4 Rain, Snow, and Ice

The system shall be designed as an outdoor installation (except those portions within enclosed structures) and shall be capable of operation during and following extreme conditions of rain, snow, and ice. Outdoor structures, other than the cooling towers, shall be designed for the loads specified in "General Criteria for Design of Civil Structures." Heat tracing or heaters or both shall be used as required. The system must furnish adequate information to the operator so that he can prevent freezing conditions within the cooling towers both at startup and during normal operation.

3.1.5 Earthquake

The system shall not be designed to withstand earthquakes.

3.2 PLANT CONDITIONS FOR DESIGN

The HR system shall be designed to function only during normal plant operating modes (e.g., system tests, startup, power generation, and normal shutdown). The system shall not be designed to operate during loss of offsite power except those components required if the CTM pumps provide a safety-related backup to the ERCW system following the design basis tornado.

3.3 IMPORTANCE TO SAFETY AND FAILURE CRITERIA

The HR system performs no safety-related function except that the CTM pumps and associated components may perform a safety-related function by furnishing a backup source of cool water to the ERCW pumping stations in case of tornado damage to the ERCW system. If all sources of offsite ac power are lost, it must be possible to manually align each set of components associated with the safety-related backup CTM supply to ERCW onto its designated, separate emergency power supply system. Physical separation between safety-related CTM components receiving power from different trains must be in accordance with the "Design Criteria for Physical Separations." The CTM supply to ERCW must be designed to furnish cooling water to at least one ERCW train per unit despite one single active failure in each unit in addition to resultant hazards from the tornado.

All HR system structures, components, and piping, including the cooling towers, shall be designed or located so that their failure will not damage any safety-related piping, equipment, or structures.

3.4 CYCLES OF CONCENTRATION

The CCW chemistry must be adjustable to control scaling and corrosion in the HR system. Control shall allow the concentration of dissolved solids in the HR system to be varied from twice the level in the makeup reservoir (i.e., from two cycles of concentration) to the optimum level for any month during the year.

3.5 SLIME, ALGAE, AND CLAM CONTROL

The system requires a single daily injection of sodium hypochlorite into the CCW for the control of slime and algae. The concentration of free chlorine residual developed must be calculated in ppm specifically, for each system.[2] Blowdown from the unit being treated must be suspended during that period, and not be resumed until the chlorine residual drops to an acceptable level for discharge. During suspension of blowdown, the makeup flow to the unit under treatment shall also be suspended if CTM does not supply RCW or diverted to the other unit. If CTM supplies RCW, the makeup flow must be continued and either a blowdown holding pond or additional cooling tower basin capacity provided to accommodate the RCW flow during suspension of blowdown. Additionally, sodium hypochlorite solution is to be injected manually at a point upstream of the CTM strainers for control of slime and algae buildup on the strainers. Backwash flow should be stopped during this injection.

CTM pump strainers are the primary means of Asiatic clam removal. Additionally, if RCW is taken from the CTM supply, a sodium hypochlorite solution shall be continuously injected into the CTM downstream of the CTM strainers during the Asiatic clam spawning season to control clam growth in the RCW system. During these periods, a free chlorine residual in ppm,[3] should be maintained at the discharge to the HR system.

---

2. Identify as a specific number in the design criteria.

3. Identify as a specific number in the design criteria.

191

3.6 DESIGN TEMPERATURE AND PRESSURE

The HR system shall be designed to supply CCW (cold water) to the main condenser and mfpt condenser within the range 35-95°F.[4] The hot water design temperature shall be a minimum of 35°F.

The design pressure of each HR system component is given in 5.0.

3.7 DESIGN LIFETIME

The design lifetime of the HR system and its components shall be 40 years, allowing for normal maintenance, including normal inspection, cleaning, and replacement of wearing parts of active components.

3.8 CODES AND STANDARDS

All HR system components shall be designed to the requirements of Quality Group E as defined in "Design Criteria for Mechanical Component Quality Groups."

3.9 QUALITY ASSURANCE

Items that perform a safety-related function shall be designed in accordance with applicable sections of the TVA quality assurance program as described in "The Office of Engineering Design and Construction (OEDC) Quality Assurance Manual for Design and Construction."

3.10 CORROSION CRITERIA

The following corrosion criteria shall be considered by the design engineer in establishing an appropriate wall thickness allowance for piping of noncorrosion resistant materials, in particular carbon steel:

Average wall reduction 0.0625 inch
Maximum local wall thinning 0.160 inch

---

4. A 95°F maximum cold water temperature is being used in this document and should always be verified or revised based on specific site conditions and restrictions.

4.0 INSTRUMENTATION AND CONTROL REQUIREMENTS

4.1 GENERAL INSTRUMENTATION AND CONTROL (I&C) REQUIREMENTS

4.1.1 Normal Operation and Control

Adequate I&C shall be installed in each unit's main control room (mcr) for the operator to monitor and operate the HR system during normal plant operation. The HR system will normally operate without requiring the operator to take any action other than manually starting and stopping the CCW, blowdown, and CTM pumps from the main control room. If RCW is supplied from CCW, makeup flow rate to the CCW shall be automatically and remote manually regulated according to the water level in the cooling tower basin or CCW pump supply with blowdown following. If RCW is supplied from CTM, makeup flow rate to the CCW shall be automatically and remote manually regulated to ensure adequate RCW supply and optimum cycles of concentration in the CCW; blowdown shall be automatically and remote manually regulated according to the water level in the cooling tower basin or CCW pump supply. Suitable provisions shall be made in the HR system design for installing the test instrumentation required to demonstrate equipment and system performance.

4.1.2 Auxiliary Control System

No HR system instrumentation and controls are required in the auxiliary control system.

4.2 SPECIFIC CONTROL REQUIREMENTS

4.2.1 CCW Pumps

Manual control shall be located locally and in the mcr so that the CCW pump motors can be manually started or stopped. In addition, the CCW pump may be started or stopped by a signal from the turbine control system.

Automatic controls shall shut off automatically the CCW pump motors upon overpressurization of the CCW system. Take particular care to avoid spurious trips in the I&C design for this function.

The CCW pump start sequence shall be synchronized with its discharge butterfly valve. Each pump shall be energized simultaneously with its discharge butterfly valve. The setpoints for this valve must be determined independently for each system; if the valve does not open a minimum number of degrees in a specific number of seconds,[5] the pump shall shut down and the valve shall close.

In the CCW pump start sequence, the bearing lubrication water (for vertical wet-pit pumps), gland seal water, and motor cooling water are being supplied, the other CCW pump discharge valves are closed or other pumps are running, and at least one path through the main condenser to the cooling tower is open before the CCW pump can be energized. The CCW pumps shall start in succession using the same starting sequence.

The normal sequence for shutting off the CCW pumps begins with closing a pump's discharge butterfly valve until the valve is in a partially open[6] position. At this point the pump shall be de-energized. The remaining CCW pumps shall be shut off in succession using the same sequence.

If power to a CCW pump is interrupted, its discharge valve shall automatically close. Precaution shall be taken to ensure that spurious signals will not close the discharge valve.

The CCW pump bearing lubrication unit and system shall be started locally. It is a vendor-designed, self-contained system. Backup RSW supply to the separator shall be automatically initiated upon low flow to the separator. If low flow to any one of the CCW pumps begins, RSW shall be automatically injected directly to the pumps. An automatic duplex strainer shall be installed downstream of the separator unit, and a means for manually injecting RSW between it and the separator.

4.2.2 Valves

Control locic for the CCW pump discharge valves is discussed in section 4.2.1.

The main condenser isolation valves shall be interlocked so that only one of the two cooling water flow paths through the main condenser can be blocked while any associated CCW pump is operating. Interlocks shall prevent any CCW pump from being started when both paths are blocked. There shall be no override on these interlocks that can defeat the functional intent, which is to prevent the HR

---

5. These setpoints must be specifically identified in the design criteria.

6. Identify a specific number of degrees in the design criteria.

system from being pressurized to the CCW pump shutoff head. Local manual control and remote manual control from the respective unit's mcr shall be available for each main condenser isolation valve and main feed pump turbine condenser isolation valve.

Control of the makeup supply to the HR system shall be accomplished by adjustment of a flow control valve (fcv) in that unit's makeup supply line (described in section 4.1.1) in conjunction with regulating the number of pumps operating. Each fcv should be capable of being operated locally and from the respective unit's mcr.

A flow control valve in the blowdown discharge line may be required to regulate water level in the cooling tower basin or CCW pump supply (see section 4.1.1).

If multiple mechanical draft cooling towers are employed, the hot water supply to each cooling tower shall incorporate a locally-controlled manual isolation valve, with the valve status indicated in the mcr. The isolation valves for a unit's cooling towers shall be interlocked to prevent any CCW pump of that unit from being started when all cooling towers for that unit are isolated. Interlocks shall prevent any CCW pump of a unit from being started when all cooling tower flow paths are blocked in the unit. There shall be no overrides on these interlocks that can defeat the functional intent, that is, to prevent the HR system from ever being pressurized to the CCW pump shutoff head.

If the CTM system has a tornado-qualified backup to the ERCW spray system, two normally open shutoff valves in series, manually-operated from the mcr, shall be incorporated into each unit's normal CTM flow path. These valves shall perform the safety-related function of diverting the CTM to the ERCW system for tornado-backup operation. The valve position shall be indicated in the mcr.

4.2.3 Cooling Towers

Local controls for cooling tower startup, normal operation, and freeze protection will be supplied by the cooling tower manufacturer. Status shall be displayed in the mcr of the cooling tower de-icing system and bypass, including the positions of all valves or gates and any other I&C that may be recommended by the cooling tower vendor for this purpose.

The cold water bypass should also be operable from the mcr.

A display in the mcr shall describe the fan operational status if mechanical draft cooling towers are employed.

4.2.4 CTM Pumps

The CTM pumps shall be manually controllable locally and from either mcr. The operational status of all CTM pumps shall be displayed in both units' mcr.

4.2.5 CTM Strainers

There shall be one CTM strainer for each CTM pump. Each strainer shall backwash continuously when its respective CTM pump is operating. Only manual, local control of each strainer backwash valve and motor shall be used. The backwash valves shall be designed for throttling.

4.2.6 Condenser Tube Cleaning System

Refer to "Design Criteria for the Condenser Tube Cleaning System."

4.2.7 Blowdown Pumps (if required)

The blowdown pumps shall be capable of being controlled manually, either locally or from the unit's mcr.

4.3 SPECIFIC INSTRUMENTATION REQUIREMENTS

4.3.1 Process Monitoring Instrumentation

4.3.1.1 Pressure

The pressure shall be monitored locally at the discharge of each CCW pump, at the discharge of each CTM pump, at the discharge of each blowdown pump (if required), and at the inlets and outlets of the main condenser and the mfpt condensers. The RCW supply pressure shall also be locally indicated.

The discharge pressure of each CCW and CTM pump and the RCW supply pressure shall be monitored in the mcr.

The differential pressure shall be monitored locally from the inlet waterbox to the intermediate waterbox and from the intermediate waterbox to the outlet waterbox of the main condenser. The differential pressure across the CTM strainer and any pressure signal used to operate the strainer backwash valve shall be monitored locally.

The differential pressure across the CCW pump bearing lube system duplex strainers shall be monitored locally.

4.3.1.2 Temperature

The temperature of the CCW at the inlet and outlet of the main condenser shall be monitored locally at the main condenser and remotely in the mcr. The temperature at the mfpt condenser inlets and outlets shall also be monitored locally.

The temperature of the cold water in the basin of each cooling tower shall be displayed in the mcr.

Operating conditions within the CCW, CTM, and blowdown pumps and motors shall be monitored in the mcr.

The temperature of the combined blowdown of the two units shall be monitored in the mcr of unit 1 and continuously recorded. The makeup water temperature shall also be monitored in the mcr of unit 1 and continuously recorded.

4.3.1.3 Flow

Each unit's makeup and blowdown flow rates shall be monitored in its respective mcr. The plant's combined blowdown rate shall be continuously recorded.

The CCW pump bearing lube water flow shall be monitored locally at the CCW supply to the separator unit and at the supply to each CCW pump.

4.3.1.4 Level

The water level in each unit's cooling tower basin or CCW pumping station forebay shall be monitored in that unit's mcr (see section 4.3.3.4).

4.3.1.5 Water Quality

Local monitoring is required for pH and suspended solids discharged from the desilting pond or holdup pond (if required).

4.3.2 Test Instrumentation

4.3.2.1 Pressure

Accommodations for pressure test instrumentation shall be installed at each CCW pump discharge. Also, accommodations for installing pressure test instrumentation shall be made at the inlet and outlet of each main condenser and mfpt condenser.

4.3.2.2 Temperature

Accommodations shall be made for installing test instrumentation to measure temperature at the inlet and the outlet of each main condenser and mfpt condenser. The instrumentation at the outlets of the main condenser shall consist of four temperature test connections installed at equal angular intervals around the circumference of the discharge, skewed at a 45-degree angle in the direction of flow on vertical sections.

4.3.2.3 Flow

The flow in the CCW conduits must be measured.

4.3.3 Alarms

The following conditions shall be alarmed in the mcr unless otherwise noted.

4.3.3.1 Pressure

- Overpressurization of the CCW system

- Low RCW supply pressure

- Low CCW pump suction (if a dry-pit pump)

- Abnormally low or high CCW pump discharge pressure

- High differential pressure across a CTM pump strainer

- Low backwash flow in a CTM pump strainer

- Abnormally low or high CTM pump discharge pressure

- High differential pressure across a CCW pump bearing lube system duplex strainer (in each unit mcr)

4.3.3.2 Temperature

- Low temperature in the cooling basin

- Abnormal temperatures in the CCW, CTM, and blowdown pump or motor bearings

- High plant blowdown temperatures (in each unit mcr)

4.3.3.3 Flow

- Low combined blowdown from both units (in the unit 1 mcr)

- High and low CCW pump gland seal water flow

- High and low CCW pump bearing lube water flow

- Low supply flow to a CCW pump bearing lube separator unit

- Initiation of RSW supply for CCW pump bearing lubrication

4.3.3.4 Level

- Low level in each cooling tower basin or CCW pumping station forebay

- High level in each cooling tower basin

- Abnormal level in the CCW pump bearing lube system reservoirs, if used (locally and in each unit mcr)

5.0 SPECIFIC DESIGN CRITERIA

Certain data is needed to prepare the design calculations for a specific design project. This basic information is described in this section. Table 2, "Design Requirements for Major HRS Components" contains a list of information about the major components for the HRS design criteria for each project. The blanks are to be filled in with information used in performing the design calculations and obtained from the design calculations (see Part II of this design guide).

5.1 COOLING TOWERS

The design temperatures of water entering and leaving the cooling towers are established beforehand by the engineer.[7] The minimum temperature (always 35°F), average temperature, and maximum temperature for cold water and hot water must be known. See footnote 4 concerning the maximum cold water temperature.

The tower basin shall be designed so that water may be completely drained from it with minimal discharge of silt into the river (i.e., through a desilting basin). The access to the silt settlement area must be large enough to accommodate a front-end loader and other silt removal equipment.

---

7. Identify as a specific number in the design criteria.

TABLE 2

Design Requirements For Major HRS Components

COOLING TOWERS	
Type	
Number Required	
Capacity per tower, gal/min	

CONDENSER CIRCULATING WATER PUMPS	
Type	
Number Required	8 (4/unit)
Rated Capacity, gal/min	
Design Total Head, Ft H_2O	
Motor bhp	
Type Motor	
Design Water Temperature, °F	35°F-

COOLING TOWER MAKEUP PUMPS	
Type	Wet Pit, Vertical Turbine, Non-pullout, Above-deck, Discharge, Electric Motor Driven
Number Required	4/plant
Rated Capacity, gal/min	
Design Total Head, ft H_2O	
Design Shut-Off Head, ft H_2O	
Motor bhp	
Type Motor	

COOLING TOWER MAKEUP STRAINERS

Quantity	4
Design Capacity	
Temperature Range, °F	
Design Duty	Continuous Backwash
Maximum Pressure Differential, lb/in²	
Backwash Flow Rate	
Motor bhp	
Strainer Opening, in.	1/32

CONDENSER EXPANSION JOINTS

	Condenser Inlet	Condenser Outlet
Size, ID, in.		
End Connections	AWWA flange	AWWA flange
Material	Rubber	Rubber
Allowable Expansion, axial, in.		
lateral, in.		
Design Temperature, minimum, °F	35°F	35°F
maximum, °F		
Design Pressure, lb/in²g		

BLOWDOWN PUMPS (if required)

Type	
Number Required	3/unit
Rated Capacity	
Design Head	
Design shutoff head	
Motor bhp	
Water Temperature Range	35°F-
Type Motor	

During cold weather startup, cold water may be diverted through a cold water bypass instead of entering the cooling towers. The bypass shall be able to pass 100 percent of the total design flow.

A de-icing system shall also be a part of the cooling tower specification and design. The design of the de-icing system will depend on the manufacturer's recommendations and standard practice.

5.2 CONDENSER CIRCULATING WATER (CCW) PUMPING UNITS

Four vertical, wet-pit or dry-pit pumps shall be installed in each CCW unit. The pump motors and pumps will be installed on an open deck, therefore the motors shall have weather-protected NEMA Type II enclosures. Normal operation is three pumps; however, the system must be able to operate for extended periods on four pumps and on less than three pumps at a higher flow rate per pump. Additional data will be entered in Table 2.

5.3 PUMPING STATIONS

5.3.1 CCW Pumping Stations

If vertical wet-pit pumps are used, each pump shall have a separate sump, using the recommendations of the Hydraulic Institute Standards for the hydraulic design. Trash racks will be installed at the station inlet to protect the pumps from large debris which may fall into the open channel. It is assumed that pump maintenance will be performed by hoisting the pump from its well and moving it to a work area. It is assumed that maintenance of vertical dry-pit pumps will be performed in-place. Suction isolation valves or stoplogs are required for this purpose.

CCW pumping stations shall be designed to withstand both the horizontal and vertical thrust produced by the pumps and the discharging water. Equipment arrangement will be shown on TVA drawings.

5.3.2 Intake Pumping Station (Intk PS)

The hydraulic design of the Intk PS sump shall conform to the Hydraulic Institute Standards. The station structure must be capable of absorbing the horizontal and vertical thrust loads resulting from the CTM pump and strainer operation.

If the CTM system provides a tornado-qualified backup to the ERCW spray system, the Intk PS will be designed to protect the equipment within the structure that is essential to the ERCW spray system against tornado winds and missiles. In this event, the CTM system may be designed to withstand the pressure transients associated with the tornado by designing the station to protect the essential equipment in it from the pressure transients.

TVA drawings shall show the Intk PS equipment layout; the "Design Criteria For Intake Pumping Station" will establish the station design requirements.

5.4 VALVES

All valves for the HR system shall be single-speed, electric-motor-operated, butterfly types. Specifications for various valve applications are shown in Table 3.

5.5 PIPE AND FITTINGS

The piping and fittings shall be designed to handle the internal pressures[8] computed by the engineer in the design calculations. Normal pressures and design pressures (in lb/in^2) are required for the CCW pump suction, condenser supply, condenser discharge, blowdown, and CTM. All steel piping routed underground shall be corrosion-protected on the outside.

Flanges shall comply with AWWA C207, "Standard for Steel Pipe Flanges for Waterworks Service--Size 4 Inches Through 144 Inches."

Openings for access and dewatering the CCW conduits shall be designed into the piping system.

The interface between the conduit risers in the powerhouse and the main condenser isolation valves shall be a bolted flange,[9] located at least 12 inches above the finished floor. Expansion joints shall be located between the main condenser isolation valves and water-boxes. These expansion joints shall be standard, single arch, spool types. They shall be capable of absorbing the axial, lateral, and angular movements induced by the relative movements between the condenser and the connecting piping without creating stresses which exceed the design limits of either the connecting piping or the condenser connections.

---

8. Identify as a specific number in the design criteria.

Table 3
Valve Specifications

Valve Use	Construction Features	Operating Time (fully open to fully closed)	Notes	Design Pressure[10]
1. CCW pump discharge	Snubbers; flanged ends	30 sec.	For vertical wet-pit pump, discharge valve stem is vertical For vertical dry-pit pump, discharge valve stem is horizontal	To be determined by the engineer, for each project
2. CCW pump suction	Snubbers; flanged ends	120 sec.	Valve stems mounted horizontally	To be determined by the engineer, for each project
3. Main condenser inlet and outlet	Snubbers; flanged ends; 4-inch valve bypass for filling main condenser waterboxes	120 ±10 sec.	-	To be determined by the engineer, for each project
4. MFPT condenser inlet and outlet	Snubbers; flanged ends	-	Outlet valve to be designed for throttling	To be determined by the engineer, for each project

---

10. Identify as a specific number for the design criteria.

The expansion joints shall be of steel-reinforced elastomeric construction and flanged in accordance with AWWA C207.

The following instrumentation connections shall be installed in the CCW conduits in the powerhouse:

	No. per Conduit	Type[11]
Pressure	2	-
Temperature wells - inlet	3	-
- outlet	6	-

In addition to these connections, tube cleaning equipment connections and personnel access hatches are necessary. There shall be access hatches in both the inlet and the outlet CCW conduits, located as far as possible from the interfaces with the MC waterboxes.

5.6 CTM PUMPS

Four CTM pumps shall be installed at the Intk PS. These shall be non-pullout, vertical, wet-pit pumps with above-deck discharges. If the CTM system has a tornado-qualified backup to the ERCW spray system, the pumps and motors may be designed to withstand the pressure effects of the design basis tornado.

5.7 CTM STRAINERS

One strainer shall be installed on the discharge of each CTM pump. Each of these shall be designed for continuous operation (when its pump is operating). The strainers shall filter out particles larger than 1/32-inch. The strainer backwash shall discharge to the Intk PS trash sluice.

The CTM strainers may also perform a safety-related function by straining water supplied to the ERCW after a loss of function of the ERCW spray system because of a tornado. The CTM strainers may be designed to withstand the pressure effects of the design basis tornado.

---

11. The type of connection must be specified for the design criteria for each project.

5.8   MAIN CONDENSER WATERBOXES

Divided inlet and outlet waterboxes shall permit operation on half
the condenser while performing maintenance on the other half. Both
inlet and outlet waterboxes shall be designed to permit access for
inspection, tube plugging, and retubing. The outlet waterbox shall
support the weight of the condenser tube cleaning system.

The waterbox design pressure is established by the engineer from the
design calculations.[12]

5.9   MAIN FEED PUMP TURBINE CONDENSER WATERBOXES

The MFPT condenser waterbox design pressure is established by the
engineer from the design calculations.[13] The waterbox design shall
permit access for inspection, tube plugging, or retubing.

5.10   HEAT REJECTION SYSTEM BLOWDOWN

The blowdown piping to the diffusers shall be sized to pass a
specific flow rate of water at a predetermined pressure at the
interface between the piping and diffusers. The range of
blowdown flows shall vary according to the engineers calculations.[14]

Interfaces are required for injecting radioactive waste into the
blowdown. See "Design Criteria for Miscellaneous Liquid Waste
Management System" for interface information. Other interface
information may be found in "Design Criteria for Equipment and
Floor Drain System."

If blowdown pumps are required, three per unit are required (two
normally operating).

---

12.   Identify the waterbox pressure with a specific number in the design
       criteria.

13.   Identify the mfpt condenser waterbox pressure with a specific number in
       the design criteria.

14.   The maximum flow rate, range of flow rate, and blowdown pressure must
       be specified in the design criteria.

5.11 COOLING TOWER MAKEUP

The CTM piping shall be sized to pass a specific maximum flow rate
water for each unit. The piping design pressure is established by
the engineer from the design calculation.[15]

In the event the ERCW spray system becomes inoperable as a result
of the design basis tornado, the CTM supply should be designed for
an alternate use as the ERCW cooling water supply. Hydraulically,
this will involve diverting the CTM water from the cooling towers
via two separate lines to the ERCW pumping stations by opening and
closing remote, manually-operated valves. If a tornado occurs
concurrently with a loss of offsite power, it will be necessary to
load certain HR system components onto their own designated,
separate, emergency power system. This would also be accomplished
manually by closing the proper circuit breakers. Assuming no
active failures, the four CTM pumps would supply water at the
rate equivalent to that for operation of all four ERCW trains.
However, two CTM pumps should be able to supply the minimum
safety-related ERCW flow rate, which is that for operation of
two ERCW trains.

5.12 CCW PUMP BEARING LUBE WATER OR SEAL WATER SYSTEM

There shall be one cyclone separator system per unit, designed to
filter out any particles greater than 75 microns in diameter. If
necessary, a pump shall be used for injecting the filtrate into
the CCW pumps. A filtrate tank, big enough to permit servicing
the CCW pumps prior to their startup shall accompany the separator
unit. The waste slurry from the cyclone separator discharge shall
be drained back into the CCW pump suction for recirculation through
the HR system.

The normal raw water supply to each separator system shall be from
that unit's CCW pump discharge header. The backup supply shall be
the raw service water system. It shall be possible to supply both
the normal and the backup water supplies to the CCW pumps, either
through the separator unit or directly to the pumps by bypassing
the unit.

A duplex strainer, which will filter all particles greater than
75 microns in diameter, shall be located at the discharge of each
unit's bearing lube system.

---

15. The maximum flow rate and design pressure will appear in the design
criteria as specific quantities.

5.13 DESILTING POND

The cooling tower desilting pond shall have the capacity of handling the volume contained in one cooling tower basin when the depth of the water at the periphery of the basin is one foot.

5.14 FIELD TESTING CRITERIA

The cooling towers shall be field tested in accordance with the latest edition of CTI Code ATC-105, "Acceptance Test Code for Water-Cooling Tower."

The main condenser waterboxes shall be hydrostatically tested in the field at a predetermined pressure[16] in accordance with Heat Exchange Institute "Standards for Steam Surface Condensers." Each joint of CCW piping shall be field-tested for leakage as the pipe is installed. CTM and blowdown piping shall be hydrotested.

5.15 SHOP TESTING CRITERIA

The mfpt condenser waterboxes shall be hydrostatically tested in the shop at a predetermined pressure[17] in accordance with Heat Exchange Institute Standards for Steam Surface Condensers.

The system pumps, valves, piping, and strainers shall be hydrostatically tested and performance tested in the shop, using the applicable standards and specifications.

5.16 MAINTENANCE AND INSPECTION CRITERIA

All components of the system except buried, submerged, or embedded features shall be accessible for inspection during station operation.

If the CTM system is used as a tornado-qualified backup to the ERCW spray system, a spare CTM pump and motor, as well as spare parts for other HR system equipment used for the safety-related backup to the ERCW, will be stored onsite.

---

16. In the design criteria, this will be the design pressure (in lb/in^2g) plus 5 lb/in^2.

17. In the design criteria, this will be the design pressure (in lb/in^2g) plus 5 lb/in^2.

6.0 REFERENCES

6.1 CODES AND STANDARDS

The applicable standards and specifications (latest editions, including addenda) of the following agencies shall apply to the design, construction, and performance of the equipment described in this document:

a. American Society of Mechanical Engineers, ASME Section VIII, Division 1.

b. ANSI B31.1, "Power Piping," ANSI/AWWA F101, Part A, "Line Shaft Vertical Turbine Pumps," American National Standards Institute.

c. ASTM A283, "Standard Specification for Low and Intermediate Tensile Strength Carbon Steel Plates of Structural Quality," American Society for Testing and Materials.

d. Hydraulic Institute, HI.

e. NEMA MG-1, "Motors and Generators," National Electrical Manufacturers Association.

f. American Water Works Association:

AWWA C301, "Standard for Prestressed Concrete Pressure Pipe, Steel Cylinder Type, for Water and Other Liquids"

AWWA C504, "Standard for Rubber-Seated Butterfly Valves"

AWWA C203, "Standard for Coal-Tar Protective Coatings and Linings for Steel Water Pipelines--Enamel and Tape--Hot Applied"

AWWA C207, Carbon Steel Flanges

g. ATC-105, "Acceptance Test Code for Water-Cooling Towers," Cooling Tower Institute.

h. Heat Exchange Institute Standards for Steam Surface Condensers.

6.2 DESIGN CRITERIA

As a minimum, the following design criteria provide interface and related information:

209

a. Design Criteria for Raw Cooling Water System

b. Design Criteria for the Condenser System

c. Design Criteria for the Essential Raw Cooling Water System

d. General Criteria for Design of Civil Structures

e. Design Criteria for Environmental Design

f. Design Criteria for Design Basis Events and Required Safety Functions

g. Design Criteria for High Pressure Fire Protection System

h. Design Criteria for Chlorination System

i. Design Criteria for Liquid Waste Management System

j. Design Criteria for Sanitary Waste Management System - Yard

k. Design Criteria for Condenser Tube Cleaning System

l. Design Criteria for General Instrumentation and Controls

m. Design Criteria for Plant Drainage

n. Design Criteria for Mechanical Component Quality Groups

o. Design Criteria for Class 1E Auxiliary Power System

p. Design Criteria for Physical Separations

q. Design Criteria for Equipment and Floor Drain System

r. Design Criteria for Intake Pumping Station

s. Design Criteria for Protection of Safety-Related Components

t. Design Criteria for Noncritical Main and Reheat Steam System

u. Design Criteria Raw Service Water System

6.3 DRAWINGS

a. Design Criteria Diagram, Heat Rejection System

b. Intake Pumping Station, General Arrangement and Civil Features

210

    c.   Condenser Circulating Water Pumping Station, General Arrangement and Civil Features

    d.   Control Logic Diagram, Heat Rejection System

    e.   Design Criteria Diagram, Essential Raw Cooling Water System

    f.   Intake Pumping Station - General Plan/Location

6.4   OTHER DOCUMENTS

The Office of Engineering Design and Construction Quality Assurance Manual for Design and Construction.

Part II: Design Calculations

1.0 SCOPE

This document is a guide for performing those calculations necessary to determine the basic design parameters to be included in the heat rejection (HR) system design criteria and component specifications. The design concept reflected herein is to be employed in conjunction with Part I of this document, "Design Criteria for Heat Rejection System," and any deviations from the design concept contained in Part I should also be reflected in the procedures contained in this document.

The recommendations contained in this design guide are intended to be consistent with good engineering practice. Therefore, if circumstances arise for which good engineering practice would call for a different design approach, these recommendations should be set aside in favor of the changes.

2.0 DEFINITIONS

2.1 Ambient Wet Bulb Temperature--The wet bulb temperature of the ambient atmosphere in the immediate vicintiy of the plant, determined at a point beyond the influence of any equipment in the plant.

2.2 Asiatic Clam--A fresh water bivalve that can block or restrict the flow in raw water systems and can cause damage to machinery, both in the larva stage and in the adult stage. (See DG-M12.1.2, "Asiatic Clam Control.")

2.3 Auxiliary Power--Net generator power output less power leaving the station, i.e. power consumed by the station.

2.4 Bearing Cooling Water--Water used to cool the thrust or guide bearings of vertical dry-pit pumps.

2.5 Bearing Lube Water--Water used to lubricate the bearings of vertical wet-pit pumps.

2.6 Blowdown--Water discharged from the system to remove a portion of the scaling and other solids from the system. (See DG-M3.1.3, "Makeup and Blowdown Rates for Evaporative Cooling Systems.")

2.7 Blowdown Conduits--The piping and valves which convey the blowdown flow from the cooling tower basin to the blowdown diffuser.

2.8 Blowdown Diffuser--The device located at the end of the blowdown conduit which disperses the blowdown into the river.

212

2.9 Blowdown Holding Pond--A reservoir for storing blowdown for limited periods during which it may not be discharged into the river.

2.10 Blowdown System--The configuration of pipes and valves that transport the blowdown from the cooling tower basin to the river.

2.11 Clam Spawning Season--That time of the year when the system inlet temperature normally exceeds 60°F. (See DG-12.1.2, "Asiatic Clam Control.")

2.12 Cold Water Temperature--The temperature of the CCW measured at any point between the cooling tower basin and the main condenser inlet.

2.13 Condenser Circulating Water--The working fluid of the HR system.

2.14 Condenser Circulation Water Conduits--Piping used to transport the CCW through the HR system.

2.15 Condenser Circulating Water Pump Suction Piping--The CCW conduit between the cooling tower basin and the CCW pumps.

2.16 Condenser Circulating Water Pumping Station--The structure which supports and houses the CCW pumping units and discharge valves.

2.17 Condenser Circulating Water Pumps--Pumps that circulate the CCW through the HR system.

2.18 Condenser Discharge Piping--The CCW conduit between the main condenser and the cooling tower.

2.19 Condenser Expansion Joints--Flexible piping connectors between the main condenser and the condenser supply and discharge piping.

2.20 Condenser Supply Piping--The CCW conduit between the CCW pumps and the main condenser.

2.21 Condenser Tube Cleaning System--The configuration of strainers, pumps, balls, ball collectors, piping, and valves designed to automatically clean the main condenser tubes.

2.22 Condenser Waterbox--An enclosure on a condenser that transfers CCW from the CCW conduit to the condenser tubes or from one condenser tube bundle to the next.

2.23 Cooling Tower--An enclosed steady-flow structure for dissipating waste heat to the environment by cooling CCW through direct contact with air.

2.24 <u>Cooling Tower Approach</u>--The difference between the cold water temperature (CWT) and the wet bulb temperature of the air entering the cooling tower.

2.25 <u>Cooling Tower Basin</u>--The reservoir of CCW located immediately under the cooling tower.

2.26 <u>Cooling Tower Bypass System</u>--The configuration of piping and valves designed to convey a portion of the CCW from the condenser discharge piping to the cooling tower basin without passing over the cooling tower fill material.

2.27 <u>Cooling Tower De-Icing System</u>--The configuration of piping or channels and valves or weirs designed to permit operation of the cooling tower during cold weather without the formation of ice on the cooling tower fill material.

2.28 <u>Cooling Tower Fill Material</u>--Material placed inside the cooling tower to facilitate direct contact between the CCW and the air.

2.29 <u>Cooling Tower Makeup Pumps</u>--Pumps that deliver makeup water from the river to the cooling tower basin.

2.30 <u>Cooling Tower Range</u>--The difference between the hot water temperature (HWT) and the cold water temperature (CWT).

2.31 <u>Cooling Tower Makeup Strainer</u>--A device that strains makeup water before it is discharged into the cooling tower basin.

2.32 <u>Cycles of Concentration</u>--The ratio of total dissolved solids concentration in the system to the total dissolved solids concentration in the makeup water. (See DG-M3.1.3, "Makeup and Blowdown Rates for Evaporative Cooling Systems.")

2.33 <u>Cyclone Separation Unit</u>--A device that filters CCW before it is used as gland seal water or bearing lube water.

2.34 <u>Drift</u>--CCW lost from the cooling tower as liquid entrained in the exhaust air.

2.35 <u>Duty</u>--Rate at which heat is to be added or rejected.

2.36 <u>Essential Raw Cooling Water (ERCW)</u>--(See DG-M6.3.3, "ERCW System")

2.37 <u>Essential Raw Cooling Water Spray System</u>--The heat dissipation device used to remove heat from the ERCW. (See DG-M3.1.4, "Analysis of Closed Cycle Ultimate Heat Sinks.")

2.38 Essential Raw Cooling Water System--The ultimate heat sink, pumps, piping, valves, etc., which supply emergency raw cooling water for heat loads within the reactor building. (See DG-M6.3.3, "ERCW System.")

2.39 Essential Raw Cooling Water Train--One of two or more independent and redundant systems that supplies ERCW. (See DG-M6.3.3, "ERCW System.")

2.40 Evaporation--Water evaporated from the CCW into the atmosphere in the process of dissipating waste heat to the environment.

2.41 Gland Seal Water--Filtered water injected into a pump shaft seal to control the leakage along a pump shaft.

2.42 Heat Load--The rate at which heat is to be dissipated by the cooling towers.

2.43 Heat Rejection--A thermodynamic process essential to all cyclic heat engines whereby the waste heat is conveyed to the ambient environment. For the Rankine cycle, waste heat is transferred to the CCW in the main condenser (increasing the temperature of the CCW as the steam is condensed) and dissipated to the environment by the cooling tower.

2.44 Heat Rejection Rates--The rate at which waste heat is conveyed to the heat rejection system.

2.45 Heat Rejection System--The configuration of condensers, pumps, pipes, valves, and cooling towers designed to transport waste heat from the turbogenerator cycle to the environment.

2.46 Heat Sink--The receiving body for waste heat. For closed cycle HR systems, the heat sink is the atmosphere.

2.47 Hot Water Temperature--The temperature of the CCW as measured at the entrance to the cooling tower.

2.48 Inlet Waterbox--The condenser waterbox located on the inlet or cold-water side of the condenser.

2.49 Intake Pumping Station--The structure located on the river and which houses the CTM pumps and strainers.

2.50 Intermediate Waterbox--The condenser waterbox located between condenser tube bundles.

2.51   Main Condenser--The closed, tube-type heat exchanger which con-
        denses the turbine exhaust steam from the main turbogenerator.

2.52   Main Feedpump Turbine Condenser--The closed, tube-type heat
        exchanger which condenses the turbine exhaust steam from the main
        feedpump turbine.

2.53   Makeup--The water required to enter the system to replace water
        lost from evaporation, blowdown, drift, and other mechanisms. (See
        DG-M3.1.3, "Makeup and Blowdown Rates for Evaporative Cooling Systems.

2.54   Makeup Conduits--The piping and valves which convey the makeup flow
        from the CTM pumps to the cooling towers.

2.55   Makeup System--The configuration of pumps, strainers, pipes, and
        valves, etc. designed to transport the makeup water from the river
        to the cooling tower.

2.56   Mechanical Draft Cooling Tower--A type of cooling tower through
        which the air movement is effected by mechanical devices.

2.57   Natural Draft Cooling Tower--A type of cooling tower through which
        the air movement is effected essentially by the difference in
        densities of the ambient and exhaust atmosphere.

2.58   Outlet Waterbox--The condenser waterbox located on the outlet or
        hot-water side of the condenser.

2.59   Plant Rise--The sum of the heat rejection rates from the main
        condenser, mfpt condensers, and RCW system (if applicable) divided
        by the CCW flow rate.

2.60   Raw Cooling Water--Water used to reject or dissipate heat from
        equipment other than steam condensers. (See DG-M6.3.2, "Raw
        Cooling Water System.")

2.61   Raw Cooling Water System--A system within a thermal power plant
        that supplies the means for dissipating or rejecting heat from
        equipment other than the steam condenser, consisting of heat
        exchangers for oil coolers, air compressors, engines, motors, air
        conditioner water chillers, etc. (See DG-M6.3.2, "Raw Cooling
        Water System.")

2.62   Raw Service Water--Raw river water that is used intermittently for
        small heat loads, cleaning water, and other miscellaneous applica-
        tions.

2.63　Raw Service Water System--The pumps, controls, piping, valves, etc., that supply raw service water.

2.64　Screen Wash--Water used to clean traveling screens by impinging water on them.

2.65　Screen Wash Pumps--Pumps designed to deliver screen wash water to the traveling screens at the required flow and pressure.

2.66　Strainer Backwash--The difference between the flow through the strainer inlet and outlet connections that is diverted through the strainer backwash connection to maintain a clean strainer.

2.67　Traveling Screen--A dynamic perforated surface which operates in a closed loop, partially submerged in the intake pumping station, to screen out solids necessary to protect the pumps in the intake pumping station.

2.68　Ultimate Heat Sink--The entire ERCW system, including all backup equipment for the various modes of operation. (See DG-M3.1.4, "Analysis of Closed Cycle Ultimate Heat Sinks.")

2.69　Vertical Dry-Pit Pump--A vertical shaft centrifugal pump located in a dry well.

2.70　Vertical Wet-Pit Pump--A vertical shaft centrifugal pump located in a wet well.

2.71　Waste Heat--Energy at a thermodynamic state from which useful work cannot be economically derived.

3.0　SYSTEM DESIGN PARAMETERS

3.1　HEAT REJECTION RATES

Heat rejection rates should be calculated based on the 100 percent (guaranteed) heat balance provided by the turbogenerator manufacturer.

3.1.1　Main Condenser

The main condenser heat rejection rate should be calculated by performing a heat balance calculation around the shell side of the main condenser.

The algebraic summation of the heat crossing the boundaries of the condenser shell yields the heat rejection rate as follows:

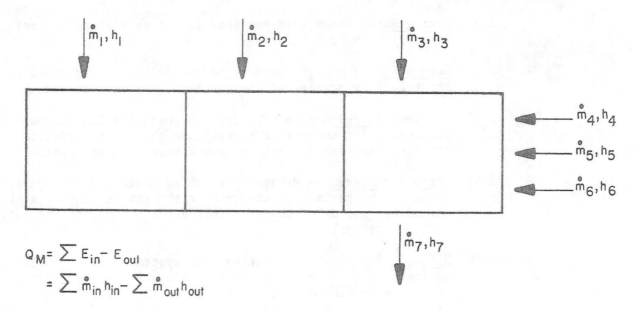

$$Q_M = \sum E_{in} - E_{out}$$
$$= \sum \dot{m}_{in} h_{in} - \sum \dot{m}_{out} h_{out}$$

Mass flows which should be considered may include, but not be limited to, turbine exhaust steam discharged into each condenser zone, mfpt condenser drains, gland steam condenser drains, heater drains, and condensate flow. The enthalpies of the various mass flows shown on the turbogenerator manufacturer's heat balance are a function of the specified main condenser back pressures. These back pressures are, in turn, a function of the sizing of the heat rejection system and the seasonal variations in the temperature of the heat sink. However, a constant heat rejection rate may be assumed for sizing the heat rejection system if the back pressures indicated on the heat balance diagram are reasonable estimates of the annual average back pressures.

3.1.2 MFPT Condenser

The mfpt condenser heat rejection rate should be calculated by performing a heat balance calculation around the shell side of the mfpt condenser as follows:

$$Q_{MFPT} = E_{in} - E_{out}$$
$$= \overset{\circ}{m}_{MFPT} h_{in} - \overset{\circ}{m}_{MFPT} h_{out}$$
$$Q_{MFPT} = \overset{\circ}{m}_{MFPT} (h_{in} - h_{out})$$

3.1.3 RCW System

The heat rejection rate for the RCW system should be calculated by tabulating all energy losses which occur in the plant and subtracting from that total all heat loads that are known to be cooled by other systems, such as the essential raw cooling water system. The energy losses may be tabulated as follows:

a. Turbogenerator mechanical losses

b. Turbogenerator electrical losses

c. Station auxiliary power.

Turbogenerator mechanical and electrical losses are tabulated on the turbogenerator manufacturer's heat balance. The station auxiliary power must be estimated based on previous plants of similar design (see EN DES-EP 3.19, "Handling and Review of Statistical Data for TVA Thermal Electric Generating Units".) Heat loads that should be subtracted from the tabulation may include, but are not limited to, power to motors in the reactor island.

This approach for determining the RCW heat load is only an approximation, based on the assumption that all the energy produced by the plant is either exported as power or rejected as heat through the RCW, HR, or ERCW systems. This approach neglects the heat lost through piping systems and the heat dissipated directly into the atmosphere through air cooling. However, the error introduced by these simplifying assumptions is acceptable for purposes of sizing the HR system.

219

For the case where RCW serves as makeup to the HR system, the RCW heat load should not be added to the main and mfpt condenser heat rejection rates to size the cooling tower because, in this case, the annual average RCW discharge temperature is normally fairly close to the annual average cold water temperature of the HR system.

3.2   FLOWS

3.2.1   CCW Flows

The CCW flow for normal, full-load operating conditions is determined as a result of the HR system economic optimization studies (see DG-M3.1.2, "Optimization of Heat Rejection System Design"). Therefore, this discussion of flows is limited to consideration of abnormal operating conditions. These flow conditions are indicated on Figure 1 as "2-pump runout," "3-pump runback," "4-pump operation," "one waterbox isolated," and ". . . bypass."

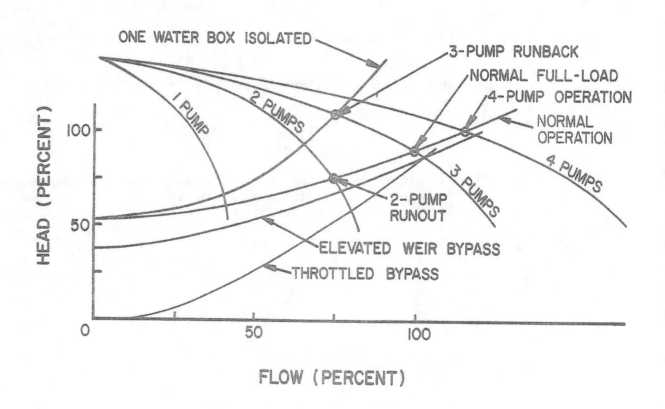

Figure 1.   Normal and Abnormal CCW Flow Conditions

The "2-pump runout" condition occurs when one of three normally operating pumps is tripped and the two remaining pumps operate on the normal system head curve. This flow is established by the intersection of the 2-pump head curve and the normal operation system head curve.

The "3-pump runback" condition is the point at which three pumps operate when one of the main condenser waterboxes is isolated for maintenance. This flow is established by determining the point at which the 3-pump head curve crosses the "one waterbox isolated" system head curve. This curve is developed from the normal system head curve. Four CCW pumps should never be operated with one waterbox isolated.

The "4-pump operation condition" is reserved only for extreme environmental conditions when additional CCW flow is required to avoid reducing power generation.

The cooling tower ". . . bypass" condition is used during cold weather to bypass the cooling tower while the CCW flow achieves operating temperature. Satisfactory operation with two to four CCW pumps must be achieved without the CCW overflowing into the cooling tower distribution system. Satisfactory operation of two and four pumps in the bypass mode may require throttling of the pump discharge valves to prevent excessive pump runout and to prevent overflowing into the tower distribution system, respectively. (One CCW pump should never be operated, except in the starting sequence, unless its discharge valve is simultaneously throttled to prevent excessive pump runout.)

The HR system should be capable of operation anywhere between flows corresponding to two pumps running, normal operation and one water box isolated system curves; three pumps running, both curves; and four pumps running, normal operation system curve. The CCW pump discharge valves will have throttling capability. If the flow through the main condenser is such that there is inadequate differential pressure across the tubes, the condenser tube cleaning system may need to be taken out of service.

3.2.2 Makeup and Blowdown Flows

Makeup and blowdown flows are related by a mass balance as follows:

$$\dot{m}_M = \dot{m}_B + \dot{m}_E + \dot{m}_D$$

where

$\dot{m}_M$ = mass flow rate of makeup

$\dot{m}_B$ = mass flow rate of blowdown

$\dot{m}_E$ = mass flow rate of evaporation

$\dot{m}_D$ = mass flow rate of drift.

The cycles of concentration (C) may be defined as

$$C = \frac{\dot{m}_M}{\dot{m}_B + \dot{m}_D}$$

See DG-M3.1.3, "Makeup and Blowdown Rates for Evaporative Cooling Systems," for procedures for calculating mass flow rates of evaporation and drift. With $\dot{m}_E$ and $\dot{m}_D$ given, the blowdown flow rate, $\dot{m}_B$, may be expressed as a function of C as follows:

$$\dot{m}_B = \frac{\dot{m}_E}{C - 1} - \dot{m}_D$$

See DG-M3.1.3, and DG-M3.5.2, "Evaluation of Scaling and Corrosion Tendencies of Water," for procedures for calculating design values of C to control scaling and corrosion in the system. However, if CTM provides flow to the RCW system which is then discharged back into the CTM system, RCW system flow requirements may dictate the design maximum value of C.

Based on site-dependent environmental limitations, it may be required to design the system to suspend blowdown flow for limited periods of time. If CTM provides flow to the RCW system and blowdown must be suspended, a blowdown holding pond or extra capacity in the cooling tower basin must be provided to store any excess flow, since the RCW system flow cannot be interrupted.

222

If CTM provides a backup supply of cooling water to the ERCW system, the maximum CTM flow may be dictated by the requirements of the ERCW system (see DG-M6.3.3, "ERCW System"). If CTM provides flow to the RCW system, the maximum blowdown flow will be dictated by the maximum RCW system demand and zero evaporation and drift. If a blowdown holding pond is required, consideration must also be given to the requirement for draining the pond through the blow-down line in addition to the system blowdown flow.

Makeup and blowdown flows are normally controlled by adjusting the number of CTM pumps operating and by installing throttling valves in the CTM lines.

3.3 TEMPERATURES

3.3.1 Cold Water

The cold water temperature (CWT) is the cooling tower discharge temperature. It is a function of the heat rejection rate, the CCW flow, the cooling tower performance, and ambient meteorology. The annual average, full-load CWT is determined as a result of the HR system economic optimization studies (see DG-M3.1.2). The normal maximum CWT is based on normal maximum ambient meteorology (see DG-M3.5.4, "Evaluation of Recorded Meteorological Data"), heat rejection rates based on the 100 percent (guaranteed) heat balance, 100 percent CCW flow, and design cooling tower per-formance.

The normal minimum CWT is based on the cooling tower bypass and de-icing features that maintain the tower discharge above freezing.

The blowdown temperature is equal to the CWT, since it discharges from the cooling tower basin.

3.3.2 Hot Water

The hot water temperature (HWT) is the cooling tower inlet tempera-ture and is equal to the CWT plus the plant rise. The plant rise equals the total plant heat rejection rate (including main condenser mfpt condenser, and RCW if applicable) divided by the CCW flow rate, based on a specific heat for the water of unity. The normal maximum HWT is equal to the normal maximum CWT plus the heat rejection rate based on the 100 percent (guaranteed) heat balance divided by the 100 percent CCW flow rate. The normal minimum HWT is equal to the normal minimum CWT, since the heat rejection rate could be zero. For these computations, the heat rejection rate is in Btu/hr and CCW flow is in lb/hr.

3.4 PRESSURES

3.4.1 CCW Pump Suction Piping

The normal pressure in the CCW pump suction piping (if applicable) would be the static head on the line that results from its location below the free surface water elevation in the cooling tower basin minus the piping friction. The design pressure in the CCW pump suction piping (if applicable) should be the maximum static head on the system when shut down and full of water.

3.4.2 Condenser Supply Piping

The normal pressure in the condenser supply piping is the static head on the line resulting from its location below the free surface water elevation in the cooling tower hot water distribution flume plus the dynamic losses through the system during normal operation. (See DG-M3.5.1, "Pressure Drop Calculations for Raw Cooling Water Piping and Fittings.")

The design pressure in the condenser supply piping should be the static head on the piping plus the dynamic losses through the system with one main condenser waterbox isolated and four CCW pumps running plus the transient overpressure. (See DG-M3.5.3, "Analysis and Control of Water Hammer in Large Diameter Piping Systems.")

The design pressure of the CCW pump discharge valve should be based on the shut-off head of the pump plus the maximum pressure in the CCW pump suction piping with the overpressure allowance from ANSI B31.1 applied.

3.4.3 Condenser Discharge Piping

The normal pressure in the condenser discharge piping would be the static head on the pipe that results from its location below the free surface water elevation in the cooling tower hot water distribution flume plus the dynamic losses through the system during normal operation (see DG-M3.5.1). The design pressure in the condenser discharge piping downstream of the condenser outlet isolation valve should be the pressure in this portion of the system during four-pump operation plus the transient overpressure (see DG-M3.5.3), except that if there is more than one cooling tower per unit, consideration should be given to the pressure while operating with one cooling tower isolated. The condenser discharge design pressure upstream of the isolation valve and the valve should have the same design pressure as the condenser supply piping, considering the one waterbox isolated condition.

224

3.4.4 CTM System Piping

The normal pressure in the CTM piping is the static head on the
pipe that results from its location below the free surface water
elevation in the cooling tower cold water elevation in the cooling
tower cold water basin plus the dynamic losses through the system
during normal operation (see DG-M3.5.1). The design pressure of
the CTM piping upstream of isolating valves should be based on
the maximum discharge pressure of the pump during shutoff with
ANSI B31.1 overpressure allowance applied. Design pressure of the
downstream piping should be based on the maximum pressure with
pump at shutoff and the overpressure allowance applied if one,
single valve can affect zero flow; otherwise, it should be the
pressure at the minimum flow condition.

3.5 ALARMS

3.5.1 Flow

3.5.1.1 Low Blowdown

Low blowdown flow should be alarmed at any time the radwaste
system is discharging into the blowdown line and the blowdown
flow falls below 1,000 times the maximum radwaste flow rate.

3.5.1.2 High and Low CCW Pump Service Water

The alarm on high or low CCW pump gland seal water or bearing
lube water should be set at 90 percent of design flow, to
indicate a flow restriction in the line or a rupture in the line
upstream of the flow orifice, and 150 percent of design flow to
indicate worn pump seals or a rupture in the line downstream of
the flow orifice.

3.5.2 Temperature

3.5.2.1 Low Cold Water

The alarm on low cold water in the cooling tower basin should be
set at the cold water temperature corresponding to 5°F below the
predicted cold water temperature with the most adverse
environmental conditions and with 50 percent of design heat
load, 40°F, or as recommended by the cooling tower manufacturer,
whichever is greater.

225

3.5.2.2 Abnormal CCW or CTM Pump or Motor Bearings

The alarm on abnormal CCW or CTM pump or motor bearing temperatures should be set based on the manufacturer's recommendations.

3.5.3 Pressure

3.5.3.1 High CCW System

The alarm on high CCW system pressure should be set equal to the condenser supply piping design pressure described in section 3.4.2.

3.5.3.2 Low RCW Supply

The alarm on low RCW system supply pressure should be adjusted by the operator based on seasonal temperature variations and the unit load point such that the minimum pressure required to supply the RCW flow requirements is alarmed (see DG-M6.3.2, "Raw Cooling Water System").

3.5.4 Level

3.5.4.1 Abnormal Cooling Tower Basin

The alarm on the cooling tower basin level should be set six inches above the maximum normal water level and one foot below the minimum normal water level or six inches below the lip of the overflow weir if the blowdown is discharged over an overflow weir.

3.5.4.2 Low CCW Pumping Station Forebay

This alarm should be set six inches above the minimum submergence required by the CCW pumps.

4.0 COMPONENT DESIGN PARAMETERS

4.1 COOLING TOWERS

4.1.1 Flow

The cooling tower design flow should equal 100 percent of the CCW flow (see section 3.2.1) divided by the number of cooling towers per unit. The cooling tower should be designed structurally and

hydraulically to withstand 120 percent of design flow to allow
for CCW pump head and system pressure drop tolerances. The
cooling tower should be capable of operating up to 120 percent
of design flow. The minimum normal flow would be determined by
the "one waterbox isolated" mode of operation (see section 3.2.1).
The de-icing system should be capable of protecting the cooling
tower from icing damage over a range of flows from 60 to 100
percent of design flow with a constant range and without opening
the bypass.

The cooling tower bypass should be capable of discharging the flow
of three CCW pumps directly into the cooling tower basin without
passing any flow over the fill material.

4.1.2 Range

The cooling tower range is equal to the maximum plant rise.
The maximum plant rise equals the maximum total plant heat
rejection rate (see section 3.1) divided by the CCW flow rate
(see section 3.2.1).

4.1.3 Approach

The design approach is determined from the HR system economic
studies (see DG-M3.1.2). For base-loaded plants, the design
approach is specified at the annual average wet bulb temperature
(see DG-M3.5.4, Evaluation of Recorded Meteorological Data) to
minimize the effects of potential errors in the performance of
the cooling tower at off-design conditions and to facilitate
the testing of the cooling tower. A design verification should
be performed to prove that the tower will operate under annual
average ambient conditions and that the tower design can meet
the required cold water temperature and thermal discharge limitations at extreme ambient conditions[1] without a reduction in plant
load.

---

1. Wet bulb temperature is exceeded less than 0.005 percent of the time
    (about 1/2 hour per year).

4.2   CCW PUMPING UNITS

4.2.1   Total Head

The design head of the CCW pumping units is the "total head" as defined in the Hydraulics Institute Standards as the algebraic difference between the total discharge head and the total suction head. The total discharge head is the sum of (1) the static head between the cooling tower hot water flume free surface elevation and the elevation of the CCW pump discharge centerline, (2) the friction losses (see DG-M3.5.1) between the CCW pump discharge and the cooling tower riser discharge, and (3) the velocity head at the CCW pump discharge. The total suction head is the difference between the static head and the friction head loss, plus the suction velocity head. The friction head loss is that between the cooling tower basin and the CCW pump suction. For vertical dry-pit pumps the static head is the difference between the cooling tower cold water basin minimum normal water level and the elevation of the CCW pump suction centerline. For vertical wet-pit pumps the suction static head is the difference between the cooling tower cold water basin minimum normal water level and the bottom of the pump suction bell.

Since achieving the CCW design flow is largely an economic rather than a technical objective (see section 4.4.2), no margin is required on the CCW pump head other than that which may be inherent in the pressure drop calculation procedures employed.

4.2.2   Flow

The design flow for the CCW pumping units is equal to the CCW flow (see section 3.2.1) divided by three (the number of CCW pumps per unit). Since the CCW flow rate is largely determined by an economic trade-off between facilities costs plus pumping power cost versus the improvement in net turbine heat rate and capability (see DG-M3.1.2), no additional margin is required on the CCW pump design flow.

In addition to operating normally at 100 percent of design flow, the CCW pumps must also operate satisfactorily at any of the points delineated in Figure 1. Secondary operating points may need to be specified to ensure adequate runback and runout capability.

The CCW pumps must be designed to withstand the backflow which may occur upon the loss of one or more of pumps.

228

4.2.3 Minimum NPSH or Submergence

Available npsh is defined for the CCW pumps as the atmospheric pressure, plus the static head between the cooling tower cold water basin minimum normal water level and the CCW pump suction centerline, minus the head loss between the cooling tower basin and the CCW pump suction, minus the vapor pressure of water at the normal maximum CWT (see section 3.3.1). For dry-pit CCW pumps the operating mode resulting in the minimum available npsh may be determined by the suction piping configuration. If the suction of each CCW pump is piped individually to the cooling tower basin, the most severe case may be the "cooling tower bypass" condition. This will result in the maximum required flow per pump and the maximum head loss between the cooling tower basin and the CCW pump suction. If the suction of the CCW pumps is taken off of a common pipe to the cooling towers, the most severe case may be for normal operation.

The available submergence for vertical wet-pit CCW pumps is the difference between the minimum normal water level in the CCW pump sump and the elevation of the bottom of the pump suction bell. The npsh requirements at flows higher than pump design flow will normally determine the required pump submergence. Vortexing and air injection considerations at pump design flow will probably not determine required submergence. However, both npsh and vortexing and air injection must be investigated.

4.2.4 Design Pressure

The CCW pump design pressure should be established by the pump manufacturer and should be greater than the sum of the shut-off head of the pump plus the static head between the cooling tower cold water basin maximum normal water level and the elevation of the CCW pump suction centerline (or bottom of the pump suction bell for vertical wet-pit pumps).

4.2.5 Design Temperatures

The CCW pump maximum design temperature should be established as the normal maximum HWT (see section 3.3.2) since the cooling tower bypass valve could be opened, exposing the CCW pumps to this temperature. The CCW pump minimum design temperature should be 35°F since the cooling tower bypass should maintain the CWT at or above 40°F.

4.2.6 Motor Size

The CCW pump motor should be sized such that the motor nameplate horsepower rating is not exceeded when the pump is operating at any point between those specified in section 4.2.2.

4.3 CTM PUMPING UNITS

4.3.1 Total Head

The design total head of the CTM pumping units is the system total static head above the level of water in the pump suction sump, plus the system piping friction head, plus one velocity head in the discharge. Depending upon the system configuration, the design head of the CTM pumps may be dictated by any one of the following operating modes:

a. Supplying makeup directly to the cooling towers

b. Supplying adequate pressure to the RCW system

c. Supplying adequate cooling water to the ERCW system in the event of a tornado.

Consideration must be given to satisfying all of the system operating modes when specifying the CTM pump head characteristics. The number of pumps available to satisfy these head requirements for the various modes of operation may vary depending upon the requirements for installed redundant capacity for each of the modes of operation.

4.3.2 Flow

The design flow for the CTM pumping units may be dictated by any of the possible modes of operation described in section 3.2.2 depending upon the head requirements and redundancy considerations for the various modes. The CTM pumps are normally specified to operate with best efficiency at or near the normal operating point. Secondary required operating points may need to be specified to ensure adequate runback and runout capability. The CTM system is hydraulically shared between two nuclear units, and the pumps should be sized such that the maximum normal flow requirements of both units may be met with one CTM pump out of service. If CTM provides a backup supply of cooling water to the ERCW system, however, the system must be divided during this mode of operation so that all flow requirements for two ERCW trains may be satisfied with only one-half of the CTM pumps available. Additional installed spare capacity is not normally provided for this case.

Care should be exercised to insure that other flow requirements such as CTM strainer backwash, screen wash, and additional cooling water flow requirements (if applicable) are included when sizing the CTM pumping units.

4.3.3 Design Pressure

The CTM pump design pressure (established by the pump manufacturer) should be greater than the pressure developed within the pump while operating at shut-off head with the river at the flood elevation corresponding to that which will cause an administrative shutdown of the nuclear plant.

4.3.4 Design Temperatures

The CTM pumping unit maximum and minimum design temperatures should be established based on the historical water temperatures in the river. (See the Environmental Report.)

4.3.5 Motor Size

The CTM pump motor should be sized such that the motor nameplate horsepower rating is not exceeded when the pump is operating at any point between the maximum required CTM flow and shut-off head.

4.3.6 Minimum NPSH or Submergence

The available submergence for vertical wet-pit CTM pumps is the difference between the minimum design level of the makeup reservoir and the elevation of the bottom of the pump suction bell. The required submergence must be based on vortexing considerations and npsh requirements at pump design and runout flows.

4.4 CTM STRAINERS

4.4.1 Flow

The design flow (inlet flow) for the CTM strainer should be selected after due consideration of the possible operating modes of its companion CTM pump. The CTM strainer should be compatible with the CTM pump flow requirements while utilizing any excess head that may be available.

The CTM strainer backwash flow should be established by the strainer manufacturer but should be specified not to exceed 6 percent of the strainer design flow.

231

4.4.2 Design Pressure

The design pressure for the CTM strainer should be the shutoff head of its companion CTM pump less the static head differential between the centerline of the strainer and the river flood elevation at which administrative shutdown of the nuclear plant is executed.

4.4.3 Design Temperatures

The design temperatures for the CTM strainer should be the same as that of the CTM pump (see section 4.3.4).

4.4.4 Minimum Backwash Differential Pressure

The minimum CTM strainer backwash differential pressure required to achieve the proper backwash flow should be determined as the difference between the minimum CTM pump discharge pressure under any mode of operation and the pressure required at the strainer backwash connection to overcome the static head and friction losses in the backwash piping to the discharge point. The minimum CTM strainer backwash differential pressure should be 20 lb/in^2.

4.5 MAIN CONDENSER CIRCULATING WATER CONDUIT EXPANSION JOINTS

4.5.1 Maximum Required Expansion

The maximum required axial and lateral expansion should be determined from information supplied by the main condenser manufacturer based on the temperature differential between the minimum main condenser installation temperature and the saturation temperature corresponding to the main condenser back pressure at which the main turbogenerator will trip off.

4.5.2 Design Pressure

The design pressure of the condenser expansion joints should be the static head differential between the bottom of the lowest joint and the maximum free surface water elevation in the cooling tower cold water basin, plus the CCW pump head at the flow corresponding to the "one waterbox isolated" condition (see section 3.2.1), minus the friction losses in the system between the cooling tower discharge and the expansion joint plus the transient overpressure (see DG-M3.5.3, "Analysis and Control of Water Hammer in Large Diameter Piping Systems").

4.5.3 Design Temperature

The maximum and minimum expansion joint design temperatures should be the normal maximum HWT (see section 3.3.2) and 35°F, respectively.

4.6 VALVES

4.6.1 Condenser Isolation

4.6.1.1 Design Pressure

The design pressure of the condenser isolation valves should be determined in the same manner as the condenser expansion joints (see section 4.5.2).

4.6.1.2 Design Temperature

The maximum and minimum condenser isolation valve design temperature should be the normal maximum HWT (see section 3.3.2) and 35°F, respectively.

4.6.1.3 Design Flow

The normal design flow is equal to 50 percent of normal system flow. Maximum normal flow is equal to the system flow with three CCW pumps running and one waterbox isolated. Maximum possible flow is the system flow with four pumps running and one waterbox isolated.

4.6.2 CCW Pump Discharge

4.6.2.1 Design Pressure

The CCW pump discharge valve design pressure should be the sum of the shut-off head of the pump plus the static head between the cooling tower cold water basin maximum normal water level and the elevation of the valve centerline.

4.6.2.2 Design Temperatures

The CCW pump discharge valve design temperatures should be the same as the CCW pump design temperatures (see section 4.2.5).

4.6.2.3 Design Flow

The normal design flow is equal to the CCW pump design flow. Maximum normal flow is the pump runout as determined in section 3.2.1. The valves must be able to close against a backflow of one CCW pump design flow and a head equal to CCW pump design head.

4.7 CONDENSER TUBE CLEANING SYSTEM

4.7.1 Design Pressure

The design pressure should be equal to that of the condenser discharge piping discussed in section 3.4.3.

4.7.2 Design Temperature

The maximum design temperature should be the normal maximum HWT (see section 3.3.2). The minimum design temperature should be 35°F.

4.7.3 Design Flow

The normal design flow is based on three CCW pumps in normal operation. The maximum normal flow should be based on four CCW pumps operating with both waterboxes open. The maximum possible flow for structural considerations is the system flow with four pumps running and one waterbox isolated.

4.8 MAIN CONDENSER WATERBOXES

4.8.1 Design Pressure

The design pressure should equal that of the condenser supply discharge piping discussed in sections 3.4.2 and 3.4.3, except that it can be reduced by elevation differences.

4.8.2 The maximum design temperature should be the normal maximum HWT (see section 3.3.2). The minimum design temperature should be 35°F.

4.9 MAIN FEED PUMP TURBINE CONDENSER PIPING AND WATERBOXES

4.9.1 Design Pressure

The pressure shall be that of the main condenser waterboxes adjusted for differences in elevation.

4.9.2 Design Temperature

The maximum design temperature should be the normal maximum HWT (see section 3.3.2). The minimum design temperature should be 35°F.

4.9.3 Pressure Drop

The differential pressure across the mfpt condenser and associated piping at the mfpt condenser design flow shall be less than that across the MC and associated piping.

5.0 REFERENCES

5.1 CODES AND STANDARDS

The applicable standards and specifications (latest editions, including addenda) of the following agencies shall apply to the engineering of the equipment described in this document:

a. Hydraulic Institute Standards

b. Heat Exchange Institute - Standards for Steam Surface Condensers.

c. Design Guides

- DG-M6.3.2   Raw Cooling Water System

- DG-M6.3.3   ERCW System

- DG-M3.5.1   Pressure Drop Calculations for Raw Cooling Water Piping and Fittings

- DG-M3.1.2   Optimization of Heat Rejection System Design

- DG-M3.1.3   Makeup and Blowdown Rates for Evaporative Cooling Systems

- DG-M3.5.2   Evaluation of Scaling and Corrosion Tendencies of Water

- DG-M3.5.3   Analysis and Control of Water Hammer in Large Diameter Piping Systems

- DG-M3.1.4   Analysis of Closed Cycle Ultimate Heat Sinks

- DG-M3.5.4   Evaluation of Recorded Meteorological Data

- DG-M12.1.2   Asiatic Clam Control

Chapter 8
Design Guide DG-M6.3.3: Essential Raw Cooling Water System

This design guide represents current TVA design philosophy for the essential raw cooling water (ERCW) system of a 3800-megawatt (thermal) (MW_t) nuclear plant with a light water-cooled reactor. It is intended to be an aid for the design of the ERCW system and not a substitute for good engineering practice. An engineer with a thorough understanding of the system should prepare the design criteria and perform the design calculations.

The information is presented in two parts:

Part I: Design Criteria--enables a design document draft to be
prepared for a specific project.

Part II: Design Calculations--sets forth the calculations required
for the basic design of the ERCW system.

When preparing a design criteria for a specific project, Part I of this design guide should be used as the basis for the draft. Part I is arranged in the same format as the final design criteria document. In some sections, the design requirements are well defined and explicitly stated; in this case the design guide may be quoted verbatim. In other sections, several design approaches may be possible or the final design may be otherwise indeterminate. In this latter case, the design guide will outline the choices and provide some guidance in making the final selection. The category in which a particular section falls will be apparent from the context.

Part I: Design Criteria

1.0 GENERAL

1.1 SCOPE

This design guide defines the functional design requirements of the essential raw cooling water (ERCW) system for a two-unit plant, establishes the overall system configuration, and supplies information necessary for the detailed design of ERCW components. This document, when combined with the ERCW design criteria diagram and the ERCW functional control logic diagram, establishes the conceptual design of the balance-of-plant (BOP) ERCW system.

Only the BOP design requirements of the ERCW system are covered in this document. If a standard reactor island (RI) design is being used, the manufacturer covers the ERCW features pertaining to the RI in his design. When applicable, reference will be made to the RI documents. The mechanical design requirements of all ERCW features are covered. The design requirements for civil and electrical features are not covered, except for instrumentation and control functions required to ensure proper operation, monitoring, and testing of the ERCW system.

1.2 ABBREVIATIONS

ACS	- Auxiliary control system
BOP	- Balance of plant
CCS	- Component cooling system
CTM	- Cooling tower makeup
EFWS	- Emergency feedwater system
ERCW	- Essential raw cooling water
HPFP	- High-pressure fire protection
HRS	- Heat rejection system
I&C	- Instrumentation and controls
IPS	- Intake pumping station
LOCA	- Loss of coolant accident
MCR	- Main control room
RI	- Reactor island
RSW	- Raw service water
SRDI	- Safety-related display instrumentation
SSE	- Safe shutdown earthquake
TCV	- Temperature control valve
UHS	- Ultimate heat sink

2.0 SYSTEM DESCRIPTION

The ERCW system consists of two separate and independent trains of cooling water for each reactor unit and provides a means of transferring and dissipating waste heat from the operating equipment (safety-related and

non-safety-related) listed in subsection 2.2. Heat rejection into the
atmosphere is accomplished by a closed-cycle ultimate heat sink (UHS) with
an evaporative cooling device. The UHS has an onsite 30-day storage
capacity. The system complies with the requirements of references 8.1a
and 8.1b.

2.1 SAFETY CLASSIFICATION

All ERCW features designed to Seismic Category I or Quality Groups A,
B, or C serve primary safety-related functions. Features designed to
Seismic Category I(L) serve secondary safety-related functions,
regardless of the quality group designation. Features designed to
Quality Group E with no seismic qualification are non-safety-related.
(See references 8.1c and 8.3gg.) A possible exception is the cooling
tower makeup (CTM) which, although Quality Group E, may be needed as
a tornado-qualified ERCW backup. This backup source of cooling will
be needed if the evaporative cooling device for the UHS is vulnerable
to tornado missiles. CTM is part of the heat rejection system and is
designed to Quality Group E. Therefore, any portions of the ERCW
system associated with this CTM supply may also be designed to
Quality Group E, even if it serves a primary safety function. Note
this exception in the design criteria document when it applies.

2.2 OPERATING CHARACTERISTICS

2.2.1 Operation

The ERCW system shall ensure a continuous flow of cooling water to
those systems and components necessary for plant safety during
normal operation, safe shutdown, or accident conditions. Each
reactor unit requires two separate, independent cooling circuits:
Unit 1 - Train A and Train B, and Unit 2 - Train A and Train B.
Both ERCW trains in each unit will normally be operating. Either
of the two trains in a unit shall be able to accomplish the design
objectives if the redundant train in that unit is inoperative.

Figure 1 shows the primary features, flow paths, and general configuration of the ERCW system. Two separate UHS reservoirs will be
incorporated in the system, each storing one half of the water
volume required for safe shutdown or accident mitigation for 30
days without makeup. The UHS reservoirs will be trained so that
one reservoir supplies Train A of each unit and the other supplies
Train B of each unit.

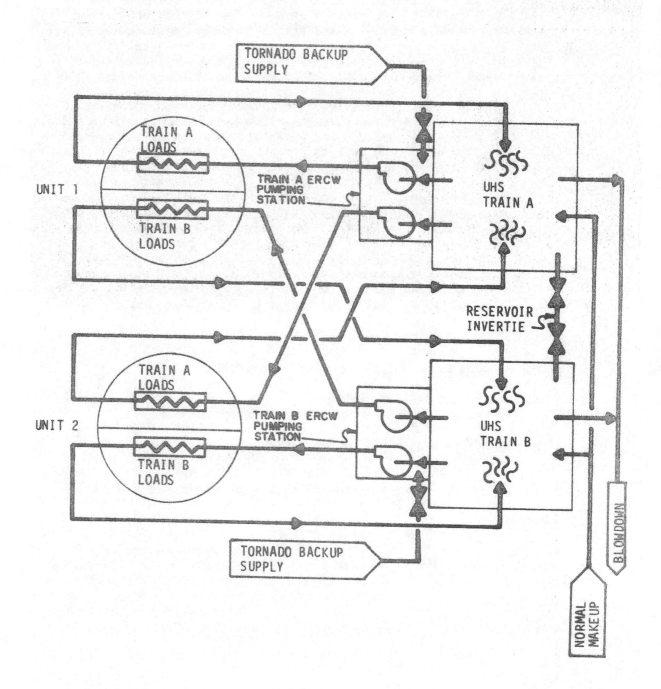

Figure 1. ERCW System Schematic

240

Each of the four ERCW trains in the plant will employ its own UHS evaporative cooling device. Each device will be able to dissipate 100 percent of the heat from the unit it serves, assuming the redundant train in that unit is inoperative. If the UHS device selected is vulnerable to impairment by tornado-generated missiles, it will be necessary to select a backup system resistant to tornado effects. CTM has been found suitable for this purpose in the past. For this application, CTM is diverted from its normal flow path to discharge into the ERCW pump suction well. From there it is pumped to the cooling loads by the ERCW pumps and discharged into the river in a once-through cooling mode. The discharge point must be carefully selected to prevent circulation back to the intake.

The concentration of dissolved solids in the water will be controlled by discharging a portion of the ERCW as blowdown. Makeup from the raw service water (RSW) system will replenish normal water losses resulting from evaporation, drift, blowdown, seepage, etc.

Water will be prevented from freezing in the UHS reservoirs, or on ERCW components that would be operationally impaired by freezing, by bypassing heated water to the UHS reservoir prior to cooling, by heat tracing, or by some other means.

2.2.2 Loads

The ERCW system shall be designed to provide:

a. Cooling for the following:
   (Itemized list to be inserted by the system engineer.)

b. Emergency makeup to the following:
   (Itemized list to be inserted by the system engineer.)

c. Emergency fire protection as indicated in reference 8.3bb.

2.3 PHYSICAL DESCRIPTION

The ERCW system will be primarily a unitized system, although a limited number of features will be shared by both units. The system components are described in the sections that follow. The quantities and types of some components vary from one system to another. During the actual design, the engineer shall decide these factors. [In this text these items are identified with an asterisk (*)].

NOTE

In this document, the phrase "manually controlled" or a similar phrase shall mean local manual control unless specifically noted otherwise.

2.3.1 Unitized Features

Following is the list of unitized ERCW components. These requirements are for one unit; requirements for the second unit are
identical.

a. Two or more* ERCW pumps per unit are needed to circulate water,
from the UHS water reservoirs to the components to be cooled,
and return the water to the UHS. The pumps shall be designated
for each cooling train, that is, Train A and Train B.

b. The UHS evaporative cooling device shall furnish adequate
cooling under all operating conditions, including the extreme
conditions described in reference 8.1b, except as noted in
subsection 2.2 for a tornado.[1] Separate UHS evaporative
cooling devices shall be installed in Trains A and B for each
unit.

c. Two bypass lines* shall be provided in each unit, one for Train
A and another for Train B. These bypass lines shall be used to
divert all or part of the heated ERCW water from the UHS
cooling device into the UHS reservoir to control reservoir
temperature and prevent freezing.

d. Additional bypass lines* shall be provided in each train to
divert the heated ERCW water to the yard drainage system (see
reference 8.3k) before reaching the UHS reservoirs when the CTM
system is being used for once-through ERCW cooling.

e. Two or more* continuous backwash strainers and appurtenances
per unit for filtration of the ERCW.

f. Instrumentation and controls (I&C) as required for the proper
operation and monitoring of the system.

g. Miscellaneous pipes, valves, and fittings.

h. One locked-closed, manual isolation valve in each ERCW line
supplying the seismically qualified portion of the
high-pressure fire protection (HPFP) system (see reference
8.3bb).

---

1. Delete the exception if the UHS evaporative cooling device is designed to
withstand tornado effects.

2.3.2 Shared Features

Those ERCW features shared by both reactor units are listed in this section. The number of features identified is that for the two-unit plant.

a. Two UHS water reservoirs, one for Train A and one for Train B. Each reservoir shall be of sufficient size to store one-half the required volume of ERCW water to permit two-unit operation for 30 days without makeup.

b. Two ERCW pumping stations for housing the ERCW pumps, screens, strainers, associated piping, valves, electrical equipment, instrumentation, and controls. One pumping station shall be designated Train A and the other Train B. Each pumping station will be at the edge of its respective reservoir.

c. Four stationary water screens, two for each reservoir, to prevent foreign objects (of a predetermined size*) from entering the ERCW pumping station sump. Each screen shall be sized to handle the maximum flow rate associated with one ERCW train in one unit.

d. One blowdown weir, together with its associated structure and piping, to discharge normal blowdown to the yard holding pond.

e. One* tornado-protected supply line from the intake pumping station to each ERCW pumping station (total of two per plant) to supply cooling water to ERCW from the CTM pumps under tornadic conditions. Each line shall branch at the ERCW pumping station with separate lines supplying each separate ERCW pump suction well. One normally closed valve shall be incorporated into each supply line within the ERCW pumping station. These valves shall be manually operated from the main control room (MCR).[2]

f. Pipes, valves, and fittings for interconnecting the Unit 1 and Unit 2 Train A ERCW supply headers, with similar accommodations for Train B. These will be locked-closed, manually operated connections and will be provided only for added flexibility.

g. Pipe, valves, and fittings for interconnecting the Train A and Train B ERCW water reservoirs. This will be a locked-closed, manually operated connection.

---

2. Delete this paragraph if CTM backup is not required.

3.0 DESIGN BASES

3.1 OPERATIONAL DESIGN BASES

The ERCW system shall be designed to dissipate the heat loads and maintain the ERCW supply temperature below the calculated design operating temperature (see subsection 3.5.1) under the following conditions:

a. Normal operation of both units.

b. Simultaneous emergency shutdown of both units.

c. Any one of the following design basis accidents in one unit with simultaneous emergency shutdown of the second unit [see reference 8.3a][3]:

    1.   Pipe break in reactor coolant pressure boundary, any size
    2.   Rupture of reactor coolant pump seals
    3.   Control element assembly ejection
    4.   Steam generator tube rupture
    5.   Inadvertent operation of pressurizer safety valve
    6.   Reactor coolant pump seizure
    7.   Steam generator depressurization inside containment
    8.   Steam generator depressurization outside containment
    9.   Destruction of main turbine
    10. Rupture of main feedwater pipe
    11. Loss of main feedwater flow
    12. Excess heat removal by steam generator
    13. Loss of one channel of vital power system
    14. Loss of one train of compressed air system
    15. Loss of main control room habitability or function
    16. Loss of trained instrument or electrical board room
    17. Rupture of waste gas decay tank
    18. Fuel handling accident inside containment
    19. Fuel handling accident outside containment
    20. Spent fuel cask drop
    21. Pipe break in another system other than ERCW

3.2 METEOROLOGICAL DESIGN BASES

The ERCW system shall be able to perform the operational design bases in subsection 3.1 during either of the following periods of adverse meteorological conditions, as required in reference 8.1b:

---

3. The engineer shall amend this list as required.

a. The 30-day period of record in the plant area resulting in the maximum use of the UHS water inventory, or

b. The single day coupled with the 29 consecutive days of record in the plant area resulting in maximum water temperatures in the UHS. The single worst day shall be inserted in the 29 consecutive worst days so that the UHS temperature is maximized.

The meteorological conditions are defined in Table 1[4] for condition a, and in Tables 2 and 3 for condition b.

3.3 NATURAL AND MISCELLANEOUS DESIGN BASIS EVENTS

The ERCW system shall be able to perform the operational design bases in subsection 3.1 with meteorological conditions described in subsection 3.2 after any one of the following design basis events. However, the design basis tornado will not be coupled with any design basis accident in subsection 3.1c that is not a postulated consequence of the tornado.

3.3.1 Design Basis Flood

The ERCW system shall be designed for continuous operation during the conditions of the design basis flood defined in reference 8.3b. (This will not require special provisions for protection of onsite ERCW components or structures since the plant grade is above the probable maximum flood level plus wave runup*.)

The ERCW system shall be designed for continuous operation during the conditions of the probable maximum precipitation. Grading in the UHS reservoir area, together with design of the yard drainage system, will prevent unacceptable erosion or debris accumulation in the vicinity of the UHS reservoirs. (See reference 8.3k.) Since the surface runoff will be below the maximum allowable ERCW intake temperature, the runoff of water into the UHS reserviors will not directly impair the cooling function of the ERCW system.

ERCW or heat rejection system (HRS) features associated with the tornado-protected CTM backup need not be protected from these flood conditions since these floods will not be coupled with the design basis tornado.[5]

---

4. Tables are at the end of the text following section 8.0.

5. Delete paragraph if CTM backup is not required.

3.3.2 Design Basis Tornado

The ERCW system shall be able to perform its design objectives (subject to the restriction in subsection 3.3 against combining the tornado with design basis accidents in subsection 3.1c) following the design basis tornado. (A design basis tornado is defined in reference 8.3b.) Requirements for protecting the ERCW stationary screens against missiles are also defined in reference 8.3b. The ERCW pumping stations shall protect all ERCW features within those structures from tornadic wind, precipitation, and missiles (see reference 8.3ii). ERCW components within the ERCW pumping stations must also be designed to withstand the effects of tornadic depressurization. All ERCW piping, valves, etc., in the yard shall also be protected from effects of the tornado in accordance with references 8.3b and 8.5c.

The design shall include a means for supplying CTM water to any or all ERCW trains simultaneously, depending on which trains are operational. Alignment of the CTM system for ERCW cooling shall be by manual main control room (MCR) operator action based on information from the cold water temperature monitors in the UHS reservoir.[6]

3.3.3 Seismic Events

All features of the ERCW system shall be designed to remain intact and functional following either the safe shutdown earthquake (SSE) or the operational basis earthquake, with the following exceptions:

a. To prevent impairment of the ERCW safety functions, features within the ERCW pumping stations associated with the CTM backup ERCW supply shall be designed to Seismic Category I(L); piping and components shall be restrained and shall maintain their pressure boundaries, and penetrations of the ERCW pumping stations shall remain leak tight following the SSE (see reference 8.3f[6]).

b. ERCW piping between the intersection with the CTM system and the ERCW pumping stations will not be required to remain intact after the SSE.[6]

c. ERCW piping associated with the CTM bypass to the yard drainage system (see subsection 2.3.1c) shall be designed to Seismic Category I(L) downstream of the normally closed isolation valve in each bypass line. Piping shall be restrained so it will not damage the seismically qualified features.[6]

---

6. Delete paragraph if CTM backup is not required.

d. ERCW blowdown piping will not be required to maintain pressure boundary after the SSE. However, it shall be restrained to protect the safety-related piping with which it might interact.

e. Features associated with the normal ERCW makeup supply from raw service water (RSW) will not be required to maintain a pressure boundary after the SSE. However, piping and components shall be restrained to preclude damage to seismically qualified features. The valves and associated piping are part of the RSW system and are discussed further in reference 8.3j. (Makeup from the RSW system will discharge into the UHS reservoirs above normal water level to prevent siphoning from the reservoirs.)

f. ERCW instrumentation and controls (I&C), other than those in Table 6, may be designed to Seismic Category I(L). The design of such I&C shall ensure that it remains intact and restrained to avoid damaging any Category I features.

With these exceptions, all other ERCW equipment, instrumentation, and controls shall be designed to Seismic Category I in accordance with reference 8.3d. The ERCW pumping stations and UHS reservoirs shall be designed in accordance with reference 8.3c.

Failure of mechanical, electrical, or civil features of systems, other than ERCW, that are not designed to Seismic Category I requirements shall not render seismically qualified ERCW features incapable of performing their intended safety functions.

3.3.4 Miscellaneous Events

The ERCW system shall remain operable after the following site-related events defined in reference 8.3a:

a. Fire
b. Explosion
c. Rupture of nonseismically qualified vessel
d. Transportation accident of land vehicle
e. Transportation accident of water vehicle

3.3.5 Combinations of Less Severe Events

The ERCW system shall remain operational following combinations of events of less severity than the foregoing design basis events. Events of this nature will be those that have historically occurred at the site or that may reasonably be expected to occur during the life of the plant.

3.4 FAILURE CRITERIA

3.4.1 Power Failure

The ERCW system shall perform the operational design basis in
subsection 3.1 with meteorological conditions described in subsection 3.2 combined with any of the natural and miscellaneous
design basis events in subsection 3.3 (subject to the restrictions
noted) concurrent with a loss of the normal sources of electrical
power. In that event, all ERCW components requiring electric power
shall be automatically loaded onto the diesel generators by the
load sequencer.

Each train of ERCW in each unit shall be powered from the corresponding train of Class 1E power. The train shall be supplied
electrical power by separate and independent conductor paths and
power sources for each unit. Cable routing and design shall ensure
independence of, and continued functional capability of, all power
trains to comply with reference 8.3g.

3.4.2 Single Failure

The ERCW system shall be designed to accomplish the operational
design bases in subsection 3.1 with meteorological conditions as
described in subsection 3.2 combined with any of the design basis
events in subsection 3.3 (subject to the restrictions noted)
combined with a single failure in one or both units. In other
words, the ERCW system shall be designed in accordance with
references 8.2a and 8.3g so that a single failure coupled with any
one of the initiating events in subsection 3.3 shall not remove
more than one train in either unit. For example, Unit 1, Train A,
and Unit 1, Train B, shall not be removed from service
simultaneously.

The ERCW mechanical single failure shall be either a single active
failure or a "limited leakage" passive failure. (In a fluid
system, a "limited leakage" passive failure is a design leak rate
based on maximum flow through a failed packing or mechanical seal
rather than based on a complete severance of the piping.) The
single electrical failures shall be those defined in reference 8.2f.

A complete loss of water from either UHS reservoir shall be considered incredible.

3.4.3 ERCW Pipe Failure

The ERCW system is classified as a moderate energy system, that is, a fluid system that has a maximum temperature of 200°F or less, and a completely severed ERCW pipe is not a design basis requirement. Rather, a "through-wall leakage crack" or "critical crack" (as defined in reference 8.3n), shall be defined as the applicable ERCW pipe failure. This pipe failure shall be considered an initiating event and is not required to be combined with any of the events in subsection 3.3.

Main control room detection of gross ERCW leakage shall be provided. These provisions are the ERCW flow and pump discharge pressure instrumentation. It is not expected that this instrumentation will be adequate to detect leakage through a critical crack. Detection of such leakage shall rely on high-level alarms for drainage sump water if the crack occurs inside a building, or by low-level alarms for UHS reservoir water if the crack occurs elsewhere. In either case, the crack must be confirmed by visual inspection.

MCR controls shall be provided to isolate a major break. These provisions are the pump and valve controls described in sub-sections 4.2.1 and 4.2.4a and d.

3.4.4 Combinations of Failures

The ERCW system shall be able to perform the operational design basis in subsection 3.1 with meteorological conditions described in subsection 3.2 combined with any of the design basis events in subsection 3.3 (subject to the restrictions noted) concurrent with both the power failure and the single failure.

The ERCW system shall also be able to perform the operational design bases in subsection 3.1 (exclusive of any pipe breaks) with meteorological conditions described in subsection 3.2 combined with both the power failure and the ERCW pipe failure.

It is not necessary to postulate a single failure concurrent with a pipe failure since, as a dual-purpose system (i.e., one that operates during both normal and accidental plant conditions), ERCW qualifies for the exemption allowed by the Nuclear Regulatory Commission (NRC) in reference 8.1e. However, it shall be a design objective to accomplish the operational design bases in sub-section 3.1 (exclusive of any pipe breaks) with meteorological conditions described in subsection 3.2 combined with both the ERCW pipe failure in subsection 3.4.3 and the single failure in sub-section 3.4.2.

3.4.5 Outage Criteria

Features shall be incorporated to permit repair or replacement of ERCW components having anticipated maintenance requirements within the allowable technical specification outage time. Assume that ERCW parts and components that may reasonably be expected to require repair or replacement over the plant life will be stored onsite for ready access.

Outage or maintenance of ERCW components in Train A or Train B of either unit shall not require a two-unit shutdown.

3.5 BASES FOR PHYSICAL DESIGN

3.5.1 Design Temperature

The maximum and minimum ERCW temperature supplied to the various cooling loads shall be specified here. A minimum temperature of 35°F is generally specified to account for cold weather startup. This can be a problem with some heating, ventilating, and air-conditioning (HVAC) users (such as chillers), possibly requiring ERCW-side flow control.

Some latitude is usually available when selecting the maximum supply temperature during the early stages of plant design. Ideally, a study would be conducted comparing cost of the UHS equipment, ERCW pumps, and user heat exchangers [particularly the Nuclear Steam Supply System (NSSS)] at different ERCW design temperatures and an optimum combination selected. Considerable margin should be allowed for state-of-the-art UHS cooling devices since NRC requirements for performance demonstration may be assumed to be tighter for such equipment.

The maximum ERCW temperature on the discharge side of each cooling load is shown in Table 5.

3.5.2 Design and Upset Pressures

The ERCW system piping may be subdivided to specify design and upset pressures, although it is seldom broken into more than two or three segments. In each segment, the design pressure selected should equal or exceed the maximum pressure in that region, assuming the system is operating normally. (Do not confuse "system" operation with "plant" operation. The ERCW "system" must operate normally during all modes of "plant" operation.)

Upset pressures shall also be specified for each section of the subdivided piping system. Upset conditions are generally associated with a control system malfunction or operator error. An example is the incorrect closure of a control valve potentially resulting in shutoff conditions upstream of the valve.

The actual value of the wall thickness allowance for corrosion that has been selected shall be clearly identified and documented as part of each piping system design calculation.

The foregoing values apply only to internal pipe corrosion in raw water systems. All piping shall be protected against external corrosion beyond normal mill scale.

Pressure drop analyses shall be in accordance with reference 8.5g.

3.6 OTHER DESIGN BASES

3.6.1 Snow, Ice, and Rain

All ERCW structures or exposed components shall withstand accumulations of snow and ice equivalent to the roof loads given in reference 8.3b. Equipment not located inside heated, qualified buildings shall be buried underground according to reference 8.3h, or an electrical freeze protection powered by the Class 1E electrical system shall be incorporated (see reference 8.3s). The UHS shall be protected from freezing as described in subsection 4.2.3.

3.6.2 Fouling and Asiatic Clam Control

The ERCW system shall be designed to minimize the effects of organic fouling. To this end, the system shall be designed to strain the ERCW and allow for the addition of a biocide into the water during normal operation without degrading the safety or reliability of the system. For control of Asiatic clams, a total chlorine residual of 0.6 to 0.8 parts per million (ppm) shall be continuously maintained in the ERCW system between the ERCW pump discharge and the inlet to the UHS reservoirs whenever the ERCW supply temperature is above 60°F. The water in any deadended headers shall be periodically renewed to prevent growth of clams in these areas (see reference 8.3p). DG-M12.1.1, "Chlorination," and DG-M12.1.2, "Asiatic Clam Control," will help with the design of this portion of the system.

3.6.3 Cycles of Concentration

For control of corrosion and scaling, the concentration of dissolved solids in the ERCW system shall normally be maintained at a level approximately twice that of the makeup reservoir (i.e., at two cycles of concentration).

Under conditions when ERCW makeup is unavailable, the cycles of concentration is expected to increase. Sufficient water shall be allocated to the UHS reservoirs to limit the cycles of concentration so that detrimental scaling does not impair the system's ability to perform its safety functions.

3.6.4 Normal Makeup

Normal makeup to the UHS reservoirs will be supplied from the RSW system, as discussed in reference 8.3j. Makeup will discharge into the UHS reservoirs above normal water level to preclude draining or siphoning the UHS reservoir in the event of failure of shutdown of the RSW system.

3.6.5 Blowdown

A weir shall be installed in each UHS reservoir to control the blowdown flow rate. The crest of the weir shall not be set below the elevation corresponding to the minimum water level required for safe shutdown.[8] The blowdown system shall be designed so that its failure will not drain water from the UHS reservoirs below the required minimum level.

3.6.6 Emergency Feedwater System Makeup[9]

The ERCW system shall provide an emergency source of makeup water to the emergency feedwater system (EFWS), as noted in reference 8.3.1. This supply shall be available for any event when the normal source of EFWS is unavailable. A predetermined maximum flow rate [in gallons per minute (gpm)*] shall be available to each reactor unit simultaneously from either ERCW train (i.e., Train A or Train B) serving that unit beginning approximately one-half hour after EFWS initiation. The water volume allocated to each UHS reservoir for this purpose is shown in Table 4. The valves and controls for establishing this flow are part of the EFWS and are described in reference 8.3(1).

3.6.7 Component Cooling System Makeup[9]

The ERCW system shall provide an emergency source of makeup water to the component cooling system (CCS), as required in reference 8.5e. This supply shall be required for any event when the normal makeup supply is available. A predetermined maximum flow rate (in gpm)* shall be available to one of the two CCS surge tanks in each unit, beginning one-half hour after shutdown initiation and extending through at least 30 days. The water volume allocated to each UHS reservoir for this purpose is shown in Table 4. The valves and controls for establishing this flow are part of the CCS and are described in reference 8.3o.

---

8. At some sites it may be necessary to pump the blowdown to the discharge point rather than by gravity flow as described here. In such cases, revise this section as required.

9. Delete this section if makeup is not required.

3.6.8 Emergency Fire Protection

The ERCW system shall be able to furnish water for fire protection*
to seismically qualified portions of the HPFP system, as noted in
reference 8.3bb. This ERCW supply will normally be isolated from
HPFP by means of locked-closed manual valves.

The water volume allocated to each UHS reservoir for emergency fire
protection is shown in Table 4.

3.6.9 CTM Supply to ERCW[10]

Features shall be incorporated in those portions of the tornado-
protected ERCW supply from the CTM system in the ERCW pumping
stations to preclude draining the UHS reservoirs in the event of
failure or misoperation of the CTM system.

4.0 INSTRUMENTATION AND CONTROL REQUIREMENTS

4.1 GENERAL INSTRUMENTATION AND CONTROL REQUIREMENTS

4.1.1 Normal Operation and Control

Adequate instrumentation and control shall be installed in the MCR
for the operator to monitor and operate the ERCW system during
normal plant operation. The design of the ERCW system shall permit
installation of instrumentation to demonstrate satisfactory
equipment and system performance.

4.1.2 Shutdown Instrumentation and Controls

Instrumentation and controls in the main control room shall enable
the operator to fully control and monitor all portions of the ERCW
system essential to safe shutdown. MCR controls shall override all
others except when plant operation has been transferred to the
auxiliary control system (ACS). Instrumentation and controls required for safe shutdown and accident mitigation shall be designed
to the applicable sections of reference 8.2e. They shall remain
functional in the applicable environmental conditions defined in
reference 8.3h, including those associated with any of the events
or conditions listed in section 3.0 for which the I&C is required
to function.

---

10. Delete paragraph if CTM backup is not required.

No reconfiguration for continued operation of the ERCW system shall be required for safe shutdown or accident mitigation other than the following:

a. The ERCW pumps and other safety-related ERCW electrical components shall be automatically sequenced onto the standby diesel generators in accordance with reference 8.5d whenever the normal sources of power are lost.

b. The CTM supply shall be manually realigned for ERCW use if the basis tornado should incapacitate the UHS cooling device. Remote manual shutoff valves will be incorporated within each ERCW pumping station for this purpose, as shown in reference 8.4a.[11]

4.1.3 Implementation

All safety-related instrumentation and controls for the ERCW system shall be implemented in accordance with reference 8.3r. Safety-related display instrumentation (SRDI) equipment shall meet the requirements specified in reference 8.3y. The SRDI parameters are tabulated in Table 6.

4.2 SPECIFIC CONTROL REQUIREMENTS

The functional control logic program for the ERCW system (see reference 8.4b) depicts those ERCW components which shall receive Class IE power by designating them Train A or Train B in their unique identification (UNID) code. Such components shall be considered as serving a primary safety function and shall be designed and qualified accordingly.

4.2.1 ERCW Pumps

Manual control of the ERCW pump motors shall be provided in the main control room, auxiliary control system, and in the ERCW pumping station. In the event that the normal sources of power are lost, the ERCW pumps shall be automatically sequenced onto the appropriate standby diesel generator in accordance with reference 8.5d.

Operational status of the ERCW pumps shall be indicated at each control station.

All ERCW pump controls are safety related.

---

11. Delete paragraph if CTM backup is not required.

4.2.2 Strainers and Backwash Valves

The ERCW strainers shall be designed for continuous backwashing
whenever their respective ERCW pump motors are running. Manual
control of the strainers and backwash valves will be required only
at the ERCW pumping station. One motor-operated backwash valve
shall be installed in each strainer backwash line. These valves
shall open to a preset, manually adjustable, intermediate position
whenever their corresponding strainer backwash motors are on.
However, they shall go to the full open position if the pressure
differential between the strainer inlet and the backwash outlet
should decrease below a calculated differential pressure,* and
shall return to the preset position when the pressure differential
increases above a calculated level.* The backwash valves shall
close automatically whenever their strainer motors are off and
shall fail in the position they were last in.

All ERCW strainer controls are safety related.

4.2.3 UHS Freeze Protection*

Upstream of the throttling valve in each train return header (see
subsection 5.2.7a), a temperature-controlled bypass shall discharge
the heated water into the UHS reservoir before it is cooled. The
bypass will reduce the cooling capability of the UHS cooling device
and prevent freezing of the reservoir in cold weather. When the
reservoir temperature drops to a specific level,* the cold water
supply temperature sensors at the ERCW pumping station shall send
control signals to the temperature control valve (TCV) in the
bypass line. This causes it to open by a preset amount, diverting
part of the heated water and degrading the effectiveness of the
UHS.

In addition to its automatic function, the TCV shall also be
manually adjustable locally and shall fail in the position it was
last in. Valve status shall be indicated locally. Controls
associated with the TCV are safety related (see subsection 5.2.7).

4.2.4 Other Valves

a. Normally open shutoff valves shall be installed at each ERCW
   penetration of primary containment. These valves shall be
   controlled locally and from the MCR and ACS. Status indication
   shall be provided at each control station. Since the ERCW is a
   "closed system inside containment," these containment isolation
   valves shall fail in the position they were last in (see
   reference 8.3q).

b. A normally closed shutoff valve shall be incorporated in each ERCW bypass line to the plant drainage system (see subsection 2.3.1c), and a normally open shutoff valve shall be installed in the ERCW flow path to the UHS reservoir downstream of the bypass connection. These valves shall be used together to divert the ERCW to plant drainage. The normally open and normally closed valves in the same unitized train must not be closed at the same time. The valves shall be manually controllable from the MCR. Status indication in the MCR is required.[12]

c. Normally closed, manually operated isolation valves shall be installed in the ERCW pumping stations for use during tornados. These valves shall be controllable from the MCR and locally, with status indication at each control station. If a tornado should render the UHS cooling device inoperative, these valves will be opened to supply cooling water to ERCW from the cooling tower makeup pumps of the heat rejection system (HRS). Valves in the normal CTM flow path will be closed at the same time to divert water to the ERCW pumping stations.[13]

d. A normally open shutoff valve shall be installed at each ERCW supply line penetration of the reactor building. These valves shall be manually controllable from the MCR with status indication in the MCR.

e. Automatic temperature control valves shall also be provided at the discharge of the coolers identified and listed at this point by the engineer.[14]

   The TCVs shall modulate the ERCW flow to these components in response to signals from process-side temperature sensors. The valves shall fail open.

   If not specifically noted, all of the foregoing automatic valves shall fail in the position they were last in. All the valve controls in subsections 4.2.4a through 4.2.4e are safety related.

   Subsections 5.2.7 and 5.2.8 contain further details about these valves.

------

12. Delete this paragraph if CTM backup is not required.

13. Delete this paragraph if CTM backup is not required.

14. All cooling loads requiring automatic temperature control valves are to be listed here.

4.3 SPECIFIC INSTRUMENTATION REQUIREMENTS

4.3.1 Process Monitoring Instrumentation

4.3.1.1 Flow

Instrumentation is required in the main control room to monitor the flow rate of the heated ERCW discharge to the UHS cooling device and in each ERCW blowdown line. Makeup flow rate to each UHS reservoir will also be monitored in the MCR.

4.3.1.2 Pressure

Instrumentation shall be provided to monitor the ERCW pressure at:

a. The discharge header from each of the ERCW pumps (local and MCR).

b. The UHS supply pressure downstream of the throttling valves described in subsection 5.2.7a (local and MCR).

c. The inlet and outlet pressure at each of the components/ systems listed in subsection 2.2.2a that require cooling (local only).

In addition, the differential pressure shall be monitored between:

d. The ERCW strainer process inlet and process outlet (local only).

e. The ERCW strainer process inlet and backwash outlet (local only).

4.3.1.3 Temperature

Instrumentation shall be provided to monitor the ERCW temperature at:

a. The cold water supply temperature on both the Unit 1 and Unit 2 sides of each ERCW pumping station suction well (MCR only).

b. The inlet and outlet temperature at each of the components/ systems listed in subsection 2.2.2a (local only). Where a particular component or system consists of more than one heat exchanger shell supplied in parallel from a common header, the supply temperature need be measured only in the common header; the discharge temperature shall be measured down-stream of each shell.

257

c. The return hot water temperature in each of the four ERCW trains (MCR only). The temperature shall be sensed in the vicinity of the flow element described in subsection 4.3.1.1, preferably upstream.

4.3.1.4 Level

Instrumentation shall be provided to monitor the entire range of water levels in each UHS reservoir upstream of the stationary water screens (MCR only). In addition, the differential level across each stationary water screen shall also be monitored (MCR only).

4.3.2 Test Instrumentation

4.3.2.1 Flow

Test instrumentation shall be provided to measure flow through each of the components/systems listed in subsection 2.2.2a. Where a particular component or system consists of more than one heat exchanger shell supplied in parallel from a common header, the flow need be measured only in the common discharge header.

Instrumentation shall also be provided to measure the blowdown flow from each ERCW reservoir.

It shall also be necessary to measure ERCW pump discharge flow rate for the ASME Section XI inservice test. If this can be done using the process monitoring instrumentation without unacceptably disturbing service to system users or requiring a significant departure from normal system alignment, no additional flow instrumentation shall be required. Otherwise, separate test instrumentation shall be added to measure flow rate from each ERCW pump.

Where the flow instrument has a significant effect on the ERCW flow rate, the instrument shall remain permanently in place. The flow monitoring instrumentation for process monitoring (subsection 4.3.1.1) may also be suitable for test purposes if it is of sufficient accuracy.

4.3.2.2 Pressure

Test instrumentation shall be supplied to measure the ERCW pump suction pressure and the UHS bypass pressure downstream of the TCV (see subsection 4.2.3).

4.3.3 Alarms

All of the following conditions shall be displayed in the main
control room. The set points for each alarm are individually
determined for each project.

4.3.3.1 Flow

An alarm shall indicate low flow at the inlet to the UHS cooling
device supply headers.*

4.3.3.2 Pressure

Pressure alarms shall be placed as follows:

a. On each ERCW pump discharge header to indicate low pressure
    when the pump has been turned on.* An appropriate time delay
    shall allow for pump startup.

b. On the ERCW strainer process inlet to process outlet
    differential pressure to indicate high differential with the
    ERCW pump running.*

4.3.3.3 Temperature

Alarms shall indicate high temperature at the following locations:

a. On the ERCW pump motor stator phases.*

b. On the ERCW pump motor bearings.*

c. On the ERCW pump bearings.*

d. On the ERCW supply to both the Unit 1 and Unit 2 sides of
    each ERCW pumping station suction wells.* Low-temperature
    alarms shall also be installed at these locations.

4.3.3.4 Level

Level alarms shall be installed as follows:

a. On the UHS reservoir water level upstream of the stationary
    screens to indicate low level.*

b. On the stationary screen differential level to indicate high
    differential.* (Differential pressure instrumentation is
    also acceptable.)

4.4 AUXILIARY CONTROL SYSTEM

Reference 8.3m states that the ERCW shall be designed to remain fully operational after any event requiring evacuation of the MCR. To this end, controls shall be provided in the ACS (see subsection 4.2). Furthermore, ACS safety-related display instrumentation shall be provided for those conditions listed.

5.0 SPECIFIC DESIGN CRITERIA

5.1 PERFORMANCE CRITERIA

5.1.1 Flow Requirements

Each train of both units shall be sized to deliver the maximum required flow rate for the operation modes in subsection 3.1. The flow rates required during normal power generation, normal shut-down, emergency shutdown, and loss of cooling accident (LOCA) shutdown are shown in Table 7. Since all ERCW trains are separate and independent, the flow requirements for any train shall not be affected by the failure of another ERCW train in either unit.

5.1.2 Heat Load Requirements

The UHS cooling device shall dissipate the maximum heat loads associated with any of the operational modes in subsection 3.1, while maintaining ERCW cold water temperatures within the safety-related design limit stipulated in subsection 3.5.1 and with meteorological conditions in subsection 3.2. The requirements of the natural and miscellaneous design basis events in subsection 3.3 and the failure criteria in subsection 3.4 shall be imposed on top of these requirements.

As shown in Table 7, the maximum heat loads occur during the loss of offsite power with simultaneous emergency shutdown of both units.

5.1.3 Inventory Requirements

Adequate water inventory shall be provided to permit operation for all four plant UHS cooling devices under the conditions of the foregoing heat loads in subsection 5.1.2 with zero makeup to the UHS reservoir for the 30-day period having the highest total water loss from evaporation, drift, seepage, etc. The meteorological conditions for this period are shown in Table 1, and the resulting water volumes to be allocated for the various modes of water loss are shown in Table 4.

The UHS reservoirs for Train A and Train B shall be interconnected, as noted in subsection 2.3.2g. Since credit will be taken for this interconnection, the total required water inventory may be divided between the two reservoirs.

It is assumed that procedures will be available for ensuring a supply of water to the reservoirs after 30 days if the normal RSW makeup supply or the tornado-qualified CTM backup supply[15] are still unavailable.

5.2 COMPONENT CRITERIA

5.2.1 ERCW Pumps

The ERCW pump (or pumps)[16] shall be supplied for each of the four ERCW trains in the plant. They shall be located in the ERCW pumping stations as shown in reference 8.4c. These horizontal, centrifugal pumps shall have a horizontally split casing for easy removal of the rotating element. They shall also have a pump/motor coupling for quick disassembly. The pump design shall meet the requirements of reference 8.2g.

Table 9 contains further details.

5.2.2 Strainers

A continuously backwashing strainer (or strainers)[17] shall be installed at the discharge of the ERCW pumps. A three-position, motor-operated valve shall be mounted in the strainer backwash line to control the backwash flow rate. The strainer backwash shall discharge to the station trash basket where the backwash debris will be separated from the water. The cleaned backwash water shall be returned to the UHS reservoir.

Table 9 contains further details.

---

15. Delete reference to CTM if not required for tornado backup.

16. The number of ERCW pumps installed per train shall be determined taking the following factors into account: technical specification allowable outage time, space restrictions, access restrictions, diesel-generated capacity limitations, separation requirements, pump supplier restraints, and cost.

17. The number of ERCW strainers per train shall be determined by taking the following factors into account: number of simultaneously operating ERCW pumps, technical specification restrictions, space restrictions, strainer supplier restraints, and cost.

5.2.3 Stationary Water Screens

There will be at least two[18] openings in each ERCW pumping station for connecting each unitized ERCW pump suction well with the UHS reservoir served by that station. A stationary water screen shall be installed across each opening. Each screen shall conform to the requirements in Table 9. Manual cleaning of the screens will be acceptable, but some means must be available to ensure that continuous screening takes place while any screen is removed for cleaning (see reference 8.3ii).

5.2.4 Ultimate Heat Sink

See reference 8.5e for the design criteria.

5.2.5 Makeup Lines

Losses from the UHS reservoirs caused by evaporation, drift, blow-down, leakage, etc., shall be replaced by makeup from the raw service water system. All features associated with this makeup supply are part of the RSW system (see references 8.3j and 8.4e). Separate makeup lines will discharge into each reservoir, although the ultimate source will be shared.

The design of the makeup system must preclude draining the UHS reservoirs under any of the conditions identified in section 3.0.

The maximum required makeup flow rate per UHS reservoir will be specified for each project* (see reference 8.3j).

5.2.6 Blowdown Lines

To maintain the desired cycles of concentration in the UHS reservoirs (see subsection 3.6.3), a portion of the ERCW shall be discharged to the yard drainage system (reference 8.3k). Blowdown from each UHS reservoir shall flow over a weir (see subsection 3.6.5). The weir shall have a crest at an elevation no lower than the specified level.[19]* Normal reservoir level is considered to be three inches above the crest of the blowdown weir.

---

18. The suction well of each ERCW pumping station will be divided into a Unit 1 side and a Unit 2 side with a concrete divider wall between them. This will allow continued operation of one unit while the suction well of the other is dewatered for maintenance. Therefore, as a minimum, two openings to the UHS reservoir will be required in each pumping station. However, more than two may be installed because of other considerations (physical size restraints, improved availability, etc.).

19. The crest height of the weir shall be selected to ensure that the water volume in Table 4 is available during a design basis event.

Blowdown from each reservoir shall be routed such that it is ultimately discharged to the yard holding pond. The blowdown flow will be specified for each project.[20]* For contingency, the piping should be designed to pass a predetermined value.[21]*

5.2.7 Modulating Valves

a. A manual throttling valve shall be incorporated in each of the ERCW return headers (total of four required per plant) for adjustment of the UHS supply flow or pressure or both.[22]

b. An automatic temperature control valve shall be incorporated in the bypass line from each of the ERCW return headers (total of four required per plant) to control the diversion of the ERCW water from the UHS cooling device into the reservoirs. The TCV shall incrementally step open to bypass water into the reservoir and thereby raise the water temperature (see subsection 4.2.3).

c. An automatic differential pressure control valve shall be provided in the backwash line from each ERCW strainer (see subsections 5.2.2. and 4.2.2).

d. Automatic temperature control valves shall be installed for the coolers listed in subsection 4.2.4e to control the process-side fluid temperatures. These valves shall incorporate features to permit local adjustment of the full-open valve position.

e. Manual throttling valves shall be incorporated at the outlet of all equipment where there are no automatic modulating valves to permit flow adjustment and balancing.

---

20. Maximum blowdown rate is that required to maintain the minimum cycles of concentration (see subsection 3.6.3) during normal operation.

21. Size to handle approximately 50 percent more than maximum expected.

22. If the required ERCW flow rate varies significantly in different operating modes, it may be necessary to incorporate an automatic pressure control valve. If so, this paragraph should be revised and a new paragraph added in subsection 4.2.4.

f. A normally closed, manual throttling valve shall be installed in a bypass line around the valve described in subsection 5.2.7a for use during maintenance of the normally open valve.

A bypass shall also be provided for the valves described in subsection 5.2.7d if it is determined that repair or replacement of these valves cannot be accomplished within the allowable ERCW technical specification outage time. However, if access to the bypass valve would not be possible within the allowable outage time, no bypass is necessary, even if the first criterion is met.

The valves discussed in subsections 5.2.7b through 5.2.7d. shall incorporate provisions to override the automatic controls, and those in subsections 5.2.7b. and 5.2.7c. shall have additional features to permit local manual control.

5.2.8 Shutoff Valves

a. A normally open shutoff valve shall be installed at each reactor building ERCW supply line penetration. These valves shall be manually controllable as discussed in subsection 4.2.4d.

b. Normally open, manually operated shutoff valves will be incorporated in the normal flow path of the CTM supply; normally closed, remote, manually operated isolation valves shall be incorporated within the ERCW pumping stations in the diverted CTM flow path to the ERCW system. In the event a tornado renders the UHS cooling system inoperative, the CTM supply will be manually aligned to provide once-through ERCW cooling by opening the valves in the diverted CTM flow path and closing those in the normal flow path. The valves in the normal CTM flow path are part of the heat rejection system and are discussed further in the design criteria document for that system (reference 8.3i). See also subsection 4.2.4c.[23]

c. In association with the CTM backup supply described in subsection 5.2.8b, a normally closed shutoff valve shall be incorporated in each ERCW bypass line to the yard drainage system (see subsection 2.3.1c). A normally open shutoff valve shall also be installed in each ERCW discharge header, downstream of the bypass connection. These valves shall be manually controllable and used as discussed in subsection 4.2.4b.[23]

---

23. Delete paragraph if CTM backup is not required.

264

d. Manual shutoff valves shall isolate each piece of equipment serviced by ERCW, including the diesel generator heat exchanger, for maintenance. These valves shall be located to isolate all instrumentation and control features associated with the equipment.

e. Normally open, manual shutoff valves shall be installed for maintenance isolation of those modulating valves discussed in subsections 5.2.7a, b, d, and e.

f. Normally open isolation valves shall be incorporated at the inlet and outlet ERCW penetrations of primary containment. These valves shall be controlled manually as discussed in subsection 4.2.4a.

g. A normally open gate valve shall be located on the suction side of each ERCW pump and shall have an extended valve stem to permit shutoff from the operating floor of the station if the pump room becomes flooded.

h. A locked-closed, manual shutoff valve shall be located on each end of the UHS reservoir interconnect line.

i. A locked-closed, manual shutoff valve shall be located on both the Unit 1 and Unit 2 sides of the ERCW pump discharge header interconnect lines. One interconnect line will be provided in each ERCW pumping station.

j. A locked-closed, manual isolation valve shall be installed in each ERCW supply line to the seismically qualified portion of the HPFP system (see reference 8.3bb).

5.3 ERCW PUMPING STATION

Each UHS reservoir requires a separate ERCW pumping station designated Train A and Train B. Unitized ERCW pump suction wells shall be provided in each pumping station.

Within the ERCW pump rooms in each pumping station, physical separation shall be provided to preclude a common hazard (e.g., fire, pipe crack) from taking both ERCW trains in the same station out of service. The design of the barrier shall allow internal flooding of the Unit 1 pump room up to the maxium water level in the UHS reservoirs without affecting Unit 2 and vice versa.

Separate drainage sumps are required in each ERCW pump room. Each sump shall be sized to store the volume of water from normal station leakage for at least 30 days without flooding. Qualified level instrumentation will be installed in each sump to alert the operator of high water level (references 8.3ff and 8.3hh). The sumps, sump pumps, and instrumentation are part of the equipment and floor drain system and are discussed in the design criteria for that system (reference 8.3dd).

These separation requirements for the ERCW pump rooms shall also be applied to the electrical equipment rooms on the operating floor, but not necessarily to the screen areas.

The design of the ERCW pumping station suction well shall comply with the recommendations of reference 8.2g.

The ERCW pumping station environmental control system (ECS) equipment shall maintain dry bulb temperature within the range specified in reference 8.3h.

6.0 MAINTENANCE, INSPECTION, AND TESTING

6.1 TESTING CRITERIA

All completed pressure boundary components and appurtenances shall be hydrostatically tested in the shop at 1.5 times their appropriate design pressure.[24] Prior to system startup, the piping system downstream of each pump suction valve shall be hydrostatically tested in the field at 1.25 times the design pressure applicable to that part of the system (see subsection 3.5.2).[24] All components, starting controls, instrumentation, and integral systems not normally operating will be functionally tested periodically.

6.2 MAINTENANCE AND INSPECTION CRITERIA

All ERCW components of Quality Group C or above shall be designed for inservice inspection in accordance with reference 8.2h. All other ERCW components shall be designed to facilitate inspection.

The ERCW system shall be designed to permit ready access to components requiring periodic replacement of wearing parts or having other preventive maintenance requirements. As noted in subsection 3.4.5, it is assumed that spare parts will be stored onsite for use when needed.

---

24. Confirm hydrostatic test requirements for the UHS cooling device.

7.0 QUALITY ASSURANCE

As a safety-related system, the ERCW system shall be designed so that it can be constructed and tested in accordance with reference 8.5a. Reference 8.5b shall be adhered to for all safety-related items which include ASME Section III components.

8.0 REFERENCES

The latest revision or edition of all reference documents shall be used unless otherwise specified.

8.1 NRC DOCUMENTS

a. NRC General Design Criteria 44 of 10CFR50, Appendix A, "Cooling Water System."

b. NRC Regulatory Guide 1.27, "Ultimate Heat Sink for Nuclear Power Plants," January 1976, Revision 2.

c. NRC Regulatory Guide 1.29, "Seismic Design Classification," September 1978, Revision 3.

d. NRC Regulatory Guide 1.48, "Design Limits and Loading Combinations for Seismic Category I Fluid System Components," May 1973.

e. NRC Branch Technical Position APCSB 3-1, "Protection Against Postulated Piping Failures in Fluid Systems Outside Containment," November 24, 1975 (attachment to Standard Review Plan SRP 3.6.1).

8.2 INDUSTRIAL STANDARDS

a. ANSI/ANS 58.9-81, "Single Failure Criteria for Light-Water Reactor Safety-Related Fluid Systems," American Nuclear Society.

b. ANSI/ANS 51.1-83, "Nuclear Safety Criteria for the Design of Stationary Pressurized Water Reactor Plants," American Nuclear Society. January 1973.[25]

---

25. Replace with ANSI/ANS 52.1-83, "Nuclear Safety Criteria for the Design of Stationary Boiling Water Reactor Plants," American Nuclear Society, if a boiling water reactor is used.

c. IEEE 308-80, "IEEE Standard Criteria for Class 1E Power Systems for Nuclear Power Generator Stations."

d. ASME Boiler and Pressure Vessel Code, Section III, "Nuclear Power Plant Components."

e. IEEE 279-71, "IEEE Standard Criteria for Protection Systems for Nuclear Power Generating Stations."

f. IEEE 379-77, "IEEE Standard Application of the Single-Failure Criterion to Nuclear Power Generating Station Class IE Systems.

g. Hydraulic Institute Standards for Centrifugal, Rotary, and Reciprocating Pumps, latest edition.

h. ASME Boiler and Pressure Vessel Code, Section XI, "Rules for Inservice Inspection of Nuclear Power Plant Components," latest revision and addendum.

8.3 DESIGN CRITERIA

The latest revision of all referenced design criteria, including Design Input Memoranda, shall be used in all cases.

a. Design Criteria for Design Basis Events and Required Safety Functions[26]

b. General Criteria for Design of Civil Structures[26]

c. Design Criteria for Dynamic Earthquake Analysis of Category I Structures and Embankments[26]

d. Design Criteria for Seismic Qualification of Category I Fluid System Components and Electrical or Mechanical Equipment[26]

e. Design Criteria for Analysis of Category I Piping System[26]

f. Design Criteria for Seismic Qualification of Category I(L) Fluid System Components and Electrical or Mechanical Equipment[26]

g. Design Criteria for Physical Separations[26]

h. Design Criteria for Environmental Design[26]

i. Design Criteria for Heat Rejection System[26]

j. Design Criteria for Raw Service Water System[26]

k. Design Criteria for Yard Drainage[26]

268

l. Design Criteria for Emergency Feedwater Systems[26]

m. Design Criteria for Safe Shutdown from Outside the Main Control Room[26]

n. Design Criteria for Evaluating and Controlling the Effects of a Pipe Failure[26]

o. Design Criteria for Component Cooling System[26]

p. Design Criteria for Chlorination System[26]

q. Design Criteria for Containment Isolation, Penetrations, and Leak Testing System[26]

r. Design Criteria for General Instrumentation and Controls[26]

s. Design Criteria for Class 1E AC Auxiliary Power System[26]

t. Design Criteria for Essential Air System[26]

u. Design Criteria for Auxiliary Area Environmental Control System[26]

v. Design Criteria for Control Building Environmental Control System[26]

w. Design Criteria for Containment Environmental Control System[26]

x. Design Criteria for Fuel Building Environmental Control System[26]

y. Design Criteria for Safety-Related Display Instrumentation[26]

z. Design Criteria for Normal AC Auxiliary Power System[26]

aa. Design Criteria for Instrumentation and Control Power System[26]

bb. Design Criteria for High Pressure Fire Protection[26]

cc. Design Criteria for Radiation Monitoring System[26]

dd. Design Criteria for Equipment and Floor Drain System[26]

ee. Design Criteria for Mechanical Component Quality Groups, and Seismic Classification[26]

ff. Design Criteria for Leakage Detection Systems[26]

---

26. Appropriate design criteria number to be filled in by engineer.

gg. Design Criteria for Essential Raw Cooling Water Pumping Stations Concrete Structures[26]

hh. Design Criteria for Protection of Safety-Related Components[26]

ii. Design Criteria for Stationary Water Screens for ERCW Pumping Stations[26]

8.4 DRAWINGS

a. ERCW Design Criteria Diagram[27]

b. ERCW Functional Control of Logic Diagram[27]

c. ERCW Pumping Station General Arrangement and Civil Features[27]

d. Heat Rejection System Design Criteria Diagram[27]

e. Raw Service Water System Design Criteria Diagram[27]

8.5 OTHER DOCUMENTS

a. The Topical Report No. TVA-TR75-A, "QA Program Description"

b. TVA Quality Assurance Manual for ASME Section III Nuclear Power Plant Components (NCM): February 14, 1975

c. Construction Specification for Earth and Rock Foundations and Fills

d. Diesel Generator Load List

e. TVA Specification for UHS Cooling System[28]

f. Electrical Design Standard, DS-E9.4.1

g. Mechanical Design Standard DS-M3.5.1, "Pressure Drop Calculations for Raw Water Piping and Fittings"

---

27. Drawing number to be filled in by engineer.

28. To be filled in by engineer.

TABLE 1

WORST 30 CONSECUTIVE DAYS FOR DRIFT LOSS[29]

Date	Average Wet Bulb, °F	Average Dry Bulb, °F	Average Dewpoint, °F	Average Wind Speed, mph

---

29. Meteorological data defining worst 30 days for drift loss to be filled in by engineer.

TABLE 2

WORST SINGLE DAY FOR THERMAL PERFORMANCE[30, 31]

Date[31]	Hour	Dewpoint, °F	Dry Bulb, °F	Wet Bulb °F	Relative Humidity, %

---

30. Meteorological data defining worst single day for thermal performance to be filled in by engineer.

31. Wind speed assumed to be zero.

TABLE 3

WORST 30 CONSECUTIVE DAYS FOR THERMAL PERFORMANCE[32,33]

Date	Maximum Dewpoint, °F	Minimum Dewpoint, °F	Maximum Dry Bulb, °F	Minimum Dry Bulb, °F	Maximum Wet Bulb, °F	Minimum Wet Bulb, °F

---

32. Meteorological data defining worst 30 days for thermal performance to be filled in by engineer.

33. Wind speed assumed to be zero during the 30 days.

273

TABLE 4

WATER INVENTORY REQUIREMENTS PER UHS RESERVOIR

Description	Water Volume (Gallons)
1. Heat dissipation by evaporation	_____
2. Solar evaporation	_____
3. Component cooling system makeup	_____
4. Emergency feedwater system makeup	_____
5. Allowance for seepage	_____
6. Allowance for sedimentation	_____
7. Allowance for water quality	_____
8. Drift losses	_____
9. Safety-related fire protection	_____
10. Subtotal	_____
11. Contingency	_____
12. Total	_____

TABLE 5

MAXIMUM TEMPERATURES AND PRESSURE DROPS

Component/System	Maximum ERCW Temperature (°F)		Maximum Differential Pressure[34]
	Inlet	Outlet	

---

34. Pressure drops shown are for the vendor-supplied coolers only and do not include losses in TVA piping.

TABLE 6

SAFETY-RELATED DISPLAY INSTRUMENTATION

Parameter or Condition	Required for Safe Shutdown (SRDI-SS)	Post-Accident Monitoring (SRDI-PAM)	Equipment Status (SRDI-ESF)
UHS reservoir water level	Yes Ref. 4.3.1.4	Yes Ref. 4.3.1.4	
UHS reservoir cold water temperature	Yes Ref. 4.3.1.3a	Yes Ref. 4.3.1.3a	
ERCW flow rate	Yes Ref. 4.3.1.1		
ERCW pump status			Yes Ref. 4.2.1
Position of all valves associated with CTW backup tornado supply[35]			Yes Ref. 4.2.4b and c
Position of containment isolation valves			Yes Ref. 4.2.4a

35. Delete if CTW tornado backup is not required.

TABLE 7

REQUIRED ERCW HEAT LOADS AND FLOWS

Component/System	Normal Power Generation[38]		Normal Shutdown[38]		Emergency Shutdown[39]		LOCA[39]	
	Flow[36]	Duty[37]	Flow[36]	Duty[37]	Flow[36]	Duty[37]	Flow[36]	Duty[37]

36. Flow per ERCW train, gpm.

37. Heat load per ERCW train, x 10^6 Btu/hr.

38. Preferred power available, one ERCW train operating in the unit.

39. Loss of preferred power, one ERCW train operating in the unit.

TABLE 8

DATA SOURCES[40]

Item No.	Component/System	Flow Rate	Heat Load	Differential Pressure

---

40. QA records for all data sources to be listed by engineer.

TABLE 9

DESIGN REQUIREMENTS FOR MAJOR ERCW COMPONENTS[41]

Pumps

Quantity	__ per ERCW train per unit
Type	Centrifugal
Design capacity	_____ gpm each
Design head	___ feet
Maximum motor horsepower rating	_____
Design shutoff pressure	____ feet
ASME code	III/3
Seismic category	I
Pump minimum available NPSH	___ feet

Strainers

Quantity	__ per ERCW train per unit
Design flow rate	_____ gpm each
Design duty	Continuous backwash
ASME	III/3
Seismic category	I
Maximum pressure differential with clean strainer elements	__ psid
Maximum backwash flow rate	____ gpm
Motor horsepower	___
Nominal size of strainer elements	1/32 inch

Stationary Water Screens

Quantity	Minimum of __ installed and __ spare per ERCW pumping station
Design flow rate	_____ gpm
Seismic category	I
Maximum differential water level (with design flow, when clean)	3 inches
Maximum opening size	____ inch

---

41. To be filled in by engineer.

Part II: Design Calculations

TABLE 9

DESIGN REQUIREMENTS FOR MAJOR ERCW COMPONENTS[41]

Pumps

Quantity	__ per ERCW train per unit
Type	Centrifugal
Design capacity	_____ gpm each
Design head	___ feet
Maximum motor horsepower rating	_____
Design shutoff pressure	_____ feet
ASME code	III/3
Seismic category	I
Pump minimum available NPSH	___ feet

Strainers

Quantity	__ per ERCW train per unit
Design flow rate	_____ gpm each
Design duty	Continuous backwash
ASME	III/3
Seismic category	I
Maximum pressure differential with clean strainer elements	__ psid
Maximum backwash flow rate	___ gpm
Motor horsepower	___
Nominal size of strainer elements	1/32 inch

Stationary Water Screens

Quantity	Minimum of __ installed and __ spare per ERCW pumping station
Design flow rate	_____ gpm
Seismic category	I
Maximum differential water level (with design flow, when clean)	3 inches
Maximum opening size	_____ inch

---

41. To be filled in by engineer.

Where C = Number of cycles of concentration

E = Maximum normal evaporation rate, gpm

L = Leakage rate, gpm

S = Seepage rate, gpm

D = Drift rate, gpm

It is recommended that the blowdown system be sized to permit operation at 2.0 cycles of concentration (C = 2). However, it is desirable to operate the cooling system at so-called "optimum" cycles of concentration to minimize scaling and corrosion. This "optimum" value may be found using techniques in Mechanical Design Guide DG-M3.5.2, "Evaluation of Scaling and Corrosion Tendencies of Water." Optimum C should be based on maximum ERCW water temperature in the return header. If all heat exchanger duties are not known, the maximum temperature may be estimated as follows:

$HWT = CWT_{design} + 25$

Where HWT = Maximum ERCW hot water temperature, °F

CWT_{design} = ERCW cold water design temperature, °F
(see Part I, subsection 3.5.1)

Evaporation results primarily from dissipation of the heat load from cooled equipment. However, solar evaporation can also be significant for certain types of cooling equipment (spray ponds, cooling ponds, etc.). Therefore, the total evaporation rate, E, is represented by the following equation:

$E = E_{HL} + E_{solar}$

Where E_{HL} = Heat load evaporation rate, gpm

E_{solar} = Solar evaporation rate, gpm

For sizing the blowdown system, it should be assumed that all heat is dissipated by evaporation alone (that is, assume zero convective heat transfer).

Guidance for calculating E_{HL} may be found in Mechanical Design Guide DG-M3.1.3, "Makeup and Blowdown Rates for Evaporative Cooling Systems." The maximum normal heat load should be used in calculating E_{HL}.

The solar heat load should ideally be computed from actual data applicable to the plantsite. If such data is not available, data from the National Climatic Center for a nearby recording station may be used (refer to Mechanical Design Guide DG-M3.5.4, "Evaluation of Recorded Meteorological Data"). Also acceptable are published statistical data from sources such as the Handbook for Air-Conditioning, Heating, and Ventilating and the Climatic Atlas of the United States (published by the Department of Commerce). In any case, a maximum average daily solar heat load should be calculated. Solar evaporation can then be calculated in the same manner as E_{HL}.

Leakage may result from losses in packing and seals, flange joints, etc. Such losses are generally quite small and may be estimated from the equation:

$$L = .0005 \times Q$$

Where Q = ERCW flow rate, gpm
L = Leakage rate, gpm

Leakage may also occur as seepage, S, through the clay liner of spray ponds, cooling ponds, etc. Such leakage can be significant. An estimate of this quantity should be obtained from the responsible civil engineer.

The final factor in the blowdown equation is drift, D. Guidance for calculating this quantity can be found in Mechanical Design Guide DG-M3.1.3.

2.2.2 Makeup

Makeup, M, is required to replace water losses from the ERCW system. The makeup system should be sized to supply the maximum anticipated flow rate during normal operation, using the following equation:

$$M = B + L + S + D + E$$

2.3 TEMPERATURES

2.3.1 Cold Water Temperature

The cold water temperature (CWT) is the ERCW temperature delivered to the system users. CWT is a function of heat load, ERCW flow rate, cooling system performance, and ambient meteorological conditions. However, selection of the design basis maximum CWT generally involves a choice of discrete options provided by the Nuclear Steam Supply System (NSSS) supplier (e.g., 90°F, 95°F, 100°F). Higher temperatures will require larger heat exchangers and higher ERCW flow rates but will place less demand on cooling system performance. Therefore, the selection of a design basis maximum CWT should involve consideration of such economic factors as the value of heat exchanger space requirements, the cost of ERCW piping, the operating cost of the ERCW pumps, and the cost of the ERCW cooling system. Intangibles, such as confidence in the ability of the cooling system to produce the selected CWT, may also be considered.

The design basis minimum is 35°F as noted in subsection 3.5.1 of Part I.

Mechanical Design Guide DG-M3.1.4 can give further guidance on CWTs.

2.3.2 Hot Water Temperature

The hot water temperature (HWT) is the ERCW temperature entering the ultimate heat sink (UHS) cooling device (cooling tower, spray system, etc.). The HWT may be calculated from the following equation:

$$HWT = CWT + \frac{HL}{500 \times Q}$$

Where CWT = ERCW cold water temperature, °F

HL = Total ERCW heat load rejected to cooling system, BTU/HR

Q = ERCW flow rate entering the UHS cooling device, gallons per minute (gpm).

This equation may also be used to find the outlet temperature from individual coolers. In this case, the HL and Q for that cooler should be used.

2.4 PRESSURES

2.4.1 Design Pressure

Article NCA-2000 of the 1977 ASME Boiler and Pressure Vessel Code gives the following definitions:

> "Design Pressure. The specified internal and external design pressure shall not be less than the maximum difference in pressure between the inside and outside of the item, or between any two chambers of a combination unit, which exists under the most severe loadings for which the Level A Service Limits are applicable. The design pressure shall include allowances for pressure surges."

"Level A Service Limits. Level A Service Limits are those sets of limits which must be satisfied for all loadings identified in the design specifications to which the component or support may be subjected in the performance of its specified service function."

The design pressure is interpreted to be the maximum pressure anticipated when the system is operating normally. Normal "system" operation does not imply normal "plant" operation for safety systems. The ERCW system is designed to operate normally for most design basis conditions considered abnormal to the plant. Therefore, the ERCW system design pressure should be based on that condition of normal system operation producing the highest internal operating pressure. Applicable variables (flow rate, water level, etc.) should be assumed at the extreme end of their normal range to maximize this value, but all valves should be in their normal configuration for the event.

Several design pressure values may be defined for a given ERCW system. For instance, the pump suction piping usually sees only the static head from the supply reservoir. Its design pressure may be the maximum static head when shut down and full of water.

Beyond the last throttling valve the system can experience only back pressures. This will consist of interface pressure at the TVA/vendor cooling system interface, static head between the interface point and the low point in this region, and dynamic piping losses (based on aged piping) from the low point to the interface. The design pressure would be the sum of these values.

A third section would be from the ERCW pump discharge isolation valve to the final throttling valve in the system. This region would experience the pump supply pressure. Its design pressure would consist of pump total head and static head from maximum reservoir elevation to the low point, less the dynamic losses from the pump to the low point. Minimum normal flow rate should be assumed. Dynamic losses should be based on new pipe.

A fourth section would be from the pump discharge nozzle to the pump discharge isolation valve. Since it is common to start a pump against a closed or partially open discharge valve, this section can experience pump shutoff conditions. Therefore its design pressure would consist of pump shutoff head and static head from maximum reservoir water elevation to the low point of the piping in this section.

These principles may be used to further subdivide the system, if it is advantageous. An allowance for pressure surges shall be applied to each section of the system for startup, shutdown, or other transients. The allowance may be determined by procedures defined in Mechanical Design Guide DG-M3.5.3, "Analysis and Control of Water Hammer in Larger-Diameter Raw Water Piping Systems." Alternatively, an allowance of 5 pounds per square inch (psi) is generally adequate.

2.4.2 Upset Pressure

Upset conditions were defined in the 1974 ASME Boiler and Pressure Vessel Code. Although the upset definition was deleted in the 1977 Code, Level B Service Limits were defined and are considered equivalent to those for upset. Level B Service limits are defined as follows:

> "Level B Service Limits. Level B Service Limits are those sets of limits which must be satisfied for all loadings identified in the design specification for which these service limits are designated. The component or support must withstand these loadings without damage requiring repair."

The upset conditions for which Level B Service Limits apply are spurious or erroneous valve alignments or improper system operation. These may be caused by control system misoperation or manual operator action.

For the pump suction piping, no upset condition exists. For the piping downstream of the final system throttling valve, erroneously opening the valve will create an upset condition. Probably the most severe upset condition results from closing the valve farthest downstream while the pump is running, potentially deadheading all piping upstream of the valve.

3.0 COMPONENT DESIGN PARAMETERS

3.1 COOLING DEVICE

Refer to DG-M3.1.4 for this information.

3.2 ERCW PUMPS

3.2.1 Flow Rate

Two trains of ERCW equipment, Train A and Train B, shall be provided in each unit. Either train shall be able to handle all

285

ERCW design basis conditions assuming complete failure of the opposite train in that unit. Therefore each pump shall be sized to deliver the following flow rate:

$$Q_{pump} = \frac{Q_{UHS} + Q_{BW}}{N_{pump}} \times 1.1$$

Where Q_{UHS} = Flow rate to the cooling device as calculated in DG-M3.1.4, gpm

Q_{BW} = Strainer backwash flow rate, gpm

N_{pump} = Number of ERCW pumps operating per train (see comments in subsection 3.3.1)

1.1 = Flow margin factor (10%)

If the strainers have not been procured prior to pump procurement, it will be necessary to approximate the strainer backwash flow rate. A value of three percent of Q_{UHS} is reasonable. A low value is desirable to minimize diesel loading, pump size, and operating cost. The assumed backwash flow rate should then be specified as a limit in the ERCW strainer specification.

3.2.2   ERCW Pump Head

The ERCW pump is procured relatively early in the plant design. It is usually necessary to estimate the pump head requirements with preliminary or incomplete information. The following equation may be used:

$$H = [h_L + h_S + h_P + h_{INT}] \times 1.10$$

Where H = Pump total head, feet of water

h_L = Dynamic head loss, feet of water

h_S = Static head loss, feet of water

h_P = Pressurization term, feet of water

h_{INT} = Cooling system required interface pressure, feet of water

1.10 = Factor for ASME Code Section XI testing and to allow for pump wear

Since the main purpose of this calculation is to size the pump for procurement purposes, the condition analyzed will be that which will maximize H. This will likely be the design basis accident [loss of coolant accident (LOCA), emergency shutdown, etc.] requiring the largest ERCW pump flow rate.

Each term in the foregoing equation will now be discussed separately.

The first term, h_L, represents the system head loss caused by flow resistance. The h_L is calculated by summing the friction and form losses sequentially through the system from the pump (including suction piping, if any) to the cooling system interface using techniques from Mechanical Design Standard DS-M3.5.1, "Pressure Drop Calculations for Raw Water Piping and Fittings." Technically, since the ERCW system supplies many coolers in parallel, it is necessary to calculate the head loss through each cooler loop separately. However, it is generally possible to select the controlling loop or loops by inspection. Typically, such a loop is a major load with a high cooler ΔP, or the greatest static loss. The dynamic loss should also include the differential level across the stationary screens at the high level alarm point, if applicable (see subsection 4.4.2). The dynamic head loss so calculated should then be multiplied by 1.10^2 to provide a 10 percent margin for flow rate uncertainity, see subsection 3.2.1. (Head loss is assumed proportional to flow squared.)

If a reactor island type of design has been selected, an alternate procedure should be used to determine h_L. In this case, h_L will consist of a balance-of-plant (BOP) part and a reactor island (RI) part. The RI part is generally covered in an NSSS interface document. The BOP part is calculated up to the RI interface point(s) using Mechanical Design Standard DS-M3.5.1 as discussed in the preceding paragraph. The total h_L is simply the sum of the BOP and RI values. RI flow rates are generally established, so that no factor need be applied for flow margin in this case.

The second term in the pump head equation, H_S, represents static head. This is the difference in elevation between the point of atmospheric pressure equilibrium on the supply side of the ERCW pumps and the elevation of the vendor's cooling system interface with TVA piping. Static head may be calculated from the following equation.

$$h_S = (E_S)_{min} - E_{INT}) \geq 0$$

Where $(E_S)_{min}$ = Minimum elevation of free water surface at pump suction, feet MSL

E_{INT} = Elevation of TVA/vendor cooling system interface, feet MSL

NOTE

Verify that h_{INT} includes the static loss between the TVA/vendor interface and the point of atmospheric discharge. It if does not, use the elevation of the atmospheric discharge in place of E_{INT}.

The third term, hp, is included to ensure that all parts of the piping systems are pressurized to at least one atmosphere (i.e., $\geq$ barometric pressure). Pressurization is desirable to avoid problems with air liberation and cavitation. It will generally not be feasible to verify positive pressurization throughout the system at the time the pumps are procured. However, this should be considered to the extent possible. Pressurization is not generally a problem on the supply side; however, it can be a problem downstream of the flow-balancing valves in each cooler loop. The pressure at any point in the system can be determined by plotting a hydraulic gradient for each cooler loop. If any portion of the gradient falls below the physical piping or component elevation, the difference represents the amount of negative gauge pressure in the system at that point. The value to be added to the pump head for pressurizing such points is:

$$h_p = E_p - E_G \quad \text{for } E_G < E_p$$

$$= 0 \quad \text{for } E_G \geq E_p$$

Where E_G = Gradient elevation, feet MSL
E_p = Physical piping or component elevation, feet MSL
h_p = Head to be added to pressurize high point, feet of water

The fourth term in the pump head equation, h_{INT}, represents the required presure head at the interface with the vendor-supplied evaporative cooling system. The h_{INT} should be specified by the cooling system contractor. It is assumed that the datum elevation for h_{INT} is the elevation of the TVA/vendor interface. The h_{INT} is also assumed to include the static loss from the interface point to the point of atmospheric discharge; the vendor should confirm this assumtion. If this static loss is not included, it should be included in the h_S term previously defined.

288

3.2.3 NPSH or Submergence

The available net positive suction head (NPSH) at the pump inlet is calculated as follows:

$$NPSH_A = E_{S_{min}} - E_{suction} - H_{vapor} - h_L$$

Where $NPSH_A$ = Net positive suction head available, feet

$E_{S_{min}}$ = Minimum free surface water elevation at pump suction, feet MSL

$E_{suction}$ = Elevation of the pump suction nozzle centerline, feet MSL

h_{vapor} = Vapor pressure of water at maximum reservoir supply temperature, feet of water absolute

h_L = Dynamic losses from the UHS reservoir to the pump inlet nozzle, feet of water

Note that $E_{S_{min}}$ is the minimum UHS water elevation at which ERCW pump operation is required. This will occur 30 days after a design basis accident, assuming that there is no makeup during that time. The h_L term should include the loss across the stationary water screen (see subsection 4.4.2). Also, h_L should be based on the maximum pump flow rate defined in the procurement specification.

For cavitation-free pump operation, the available NPSH must equal or exceed the required NPSH specified by the vendor at maximum specified flow rate.

If a vertical, wet-pit pump is used, submergence is generally more critical than NPSH. Submergence is defined as follows:

$$S = E_{S_{min}} - E_{bell}$$

Where S = Submergence, feet of water

E_{bell} = Elevation of the bottom of the pump suction bell, feet MSL

Guidance for suction well design is given in the Hydraulic Institute Standards. As for NPSH, the available submergence must equal or exceed the value required by the vendor for vortex-free pump operation at maximum specified pump flow rate.

3.2.4 Design/Upset Pressure

The design pressure for the ERCW pumps should be calculated as follows:

$$P_{design} = [\; H_{SO} + (E_S)_{max.} - E_{pump})_{out}\;)]\; \frac{1}{2.31} + P_{surge}$$

Where P_{design} = Pump design pressure, pounds per square inch gage (psig)

H_{SO} = Pump shutoff head, feet of water

P_{surge} = Surge pressure allowance (see subsection 2.4.1), psi

$$P_{design} = [\; H_{SO} + (E_S)_{max.} - E_{discharge})\;)]\; \frac{1}{2.31}$$

Where P_{design} = Pump design pressure, psig

H_{SO} = Pump shutoff head, feet of water

$(E_S)_{max.}$ = Maximum normal pump suction reservoir free surface water elevation, feet MSL

$E_{discharge\;out}$ = Centerline elevation of pump discharge nozzle

3.2.5 Design Temperature

The design temperature for the ERCW pumps will be the same as the maximum cooling system cold water temperature (see subsection 2.3.1).

3.2.6 Motor Size

The motor nameplate horsepower rating of the ERCW pump motor must not be exceeded when the pump is operating at any point between the maximum required for ERCW flow and shutoff head.

Emergency diesel generator capacity should also be considered in the initial pump and motor sizing since this may impose restrictions.

3.3 ERCW STRAINERS

3.3.1 Flow

The ERCW strainer inlet flow should be calculated as follows:

$$Q_{strainer} = \frac{N_{pump} \times Q_{pump}}{N_{strainer}}$$

Where $Q_{strainer}$ = Inlet flow rate per strainer, gpm

Q_{pump} = Outlet flow rate per ERCW pump (from subsection 3.2.1), gpm

N_{pump} = Number of ERCW pumps operating per train

$N_{strainer}$ = Number of ERCW strainers operating per train

The number of ERCW pumps and their design flow rate have been discussed in subsection 3.2; therefore, the key unknown affecting strainer flow rate is $N_{strainer}$. Many strainer/pump arrangements are possible. The optimum arrangement should be determined from a design study considering the following factors:

a. Range of ERCW flow rate

b. Strainer backwash flow rate

c. Maintenance

d. Cost

Since strainer backwash flow rate is an additive to the ERCW pump capacity, strainer procurement should be scheduled prior to that for the ERCW pumps so that actual backwash flow rate may be used in pump sizing. If this is not possible, a backwash value will have to be assumed when determining pump capacity. This value should then be included as an upper unit in the strainer specification. A value of three percent of the strainer inlet flow is generally achievable, although potential suppliers should be surveyed for input.

3.3.2 Differential Pressure from Strainer Inlet to Strainer Outlet

The differential pressure from the strainer inlet-to-outlet nozzle is a significant contributor to the required ERCW pump head. This is another reason for scheduling strainer procurement prior to that of the pump so that guaranteed values may be used in determining pump head. If this is not feasible, a differential pressure will have to be assumed in pump sizing and included in the strainer specification as an upper limit. A value of 5 psi is generally adequate, although potential suppliers should be surveyed for input.

3.3.3 Minimum Differential Pressure from Strainer Inlet to Backwash Outlet

The minimum ERCW strainer backwash differential pressure required to achieve the proper backwash flow should be specified by the strainer manufacturer. A minimum value of 20 psi is typical.

Because of the strainer's proximity to the ERCW pumps, considerable pressure exists at the backwash connection. Conversely, since the backwash piping is generally short and discharges to atmosphere, the pressure downstream of the backwash flow control valve is usually low. The combination of high pressure drop and low back-pressure are highly conducive to cavitation at the backwash valve and in the downstream piping. Erosion resulting from high flow velocities may also be a problem. Refer to subsection 5.2 for guidance in evaluating this phenomenon.

3.3.4 Design/Upset Pressure

The design pressure for the ERCW strainers should be calculated as follows:

$$P_{design} = [H + (E_S)_{max.} - E_{strainer}) - h_L] \frac{1}{2.31} + P_{surge}$$

Where P_{design} = Strainer design pressure, psig

H = Pump total head at minimum normal flow rate, feet of water

$E_{S_{max}}$ = Maximum normal pump suction reservoir free surface water elevation, feet MSL

$E_{strainer}$ = Centerline elevation of strainer inlet nozzle, feet MSL

h_L = Dynamic head loss between pump and strainer, feet of water

P_{surge} = Surge pressure allowance (see subsection 2.4.1) psi

Upset pressure is based on pump shutoff conditions:

$$P_{upset} = [h_{SO} + (E_S)_{max.} - E_{strainer})] \frac{1}{2.31}$$

Where P_{upset} = Strainer upset pressure, psig

   h_{SO} = Pump shutoff head, feet of water

Note that the pump discharge isolation valve was assumed to be located upstream of the strainer when the foregoing equations were developed. If the valve is actually located downstream of the strainer, the foregoing equation for upset pressure should be used to calculate strainer design pressure.

3.3.5 Design Temperature

ERCW strainer design temperature should be the maximum cooling system cold water temperature. (See subsection 2.3.1.)

4.0 ALARM SET POINTS

4.1 LOW FLOW RATE

The flow alarm to indicate the ERCW return flow to the UHS cooling device of each train should be based on the minimum required ERCW flow rate during normal operation. This may result in a nuisance alarm during an accident, but will alert the operator to take corrective action to the event.

4.2 PRESSURES

4.2.1 Pump Discharge

The ERCW pump low-pressure alarm, provided to indicate low discharge pressure when the pump has been selected on, should be based on maximum normal ERCW pump flow rate, minimum normal UHS reservoir elevation, and maximum allowable pump wear (usually 10 percent, to comply with the ASME Code, Section XI).

4.2.2 Strainer Differential Pressure

The intent of the strainer differential pressure alarm is to alert the operator if the continuous backwash system is failing to keep the strainer clean. Appropriate corrective action may then be taken. Therefore, the alarm point should be based on normal operating conditions.

Strainer ΔP is quite sensitive to flow rate, usually being assumed proportional to Q^2. Conversely, ΔP is generally quite insensitive to clogging until a relatively high percentage of flow area is obstructed. Therefore, if normal flow rates can vary over a large range, detection of clogging while avoiding nuisance alarms will be difficult.

It is recommended that the vendor-supplied "% Clogged vs. ΔP Increase" curve be used to detect 75 percent clogging under maximum normal flow conditions. If there is considerable variation in the normal flow rate, an intermediate flow rate may be selected as the baseline.

The design should permit field adjustment of the ΔP set point if experience warrants a change.

4.3 TEMPERATURE

4.3.1 High ERCW System Supply Temperature

The alarm which indicates high UHS supply temperature to the ERCW pumps during normal operation is provided to detect failure or inadequate performance of the UHS cooling system or temperature-controlled bypass. The set point shall be 5°F above the expected UHS cooling device outlet temperature when operating at the 1 percent summertime wet bulb/dry bulb temperature during normal plant operation.

4.3.2 Low ERCW System Supply Temperature

An alarm shall be provided to indicate potential freezing within the UHS cooling device or reservoir. The alarm point shall correspond to the control set point for maximum ERCW bypass flow to the UHS reservoir.

4.3.3 High ERCW Pump Motor Stator Temperature

This set point should be based on vendor recommendations.

4.3.4 High ERCW Pump Motor Bearing Temperature

This set point should be based on vendor recommendations.

4.3.5 High ERCW Pump Bearing Temperature

This set point should be based on vendor recommendations.

4.4 WATER LEVEL

4.4.1 Low UHS Reservoir Level

The low UHS reservoir level is provided to indicate inadequate normal makeup or excessive loss of inventory from the UHS reservoir. The set point should be approximately three inches below the minimum normal reservoir water elevation.

4.4.2 High Stationary Screen Differential Level

Both a high and a high-high stationary-screen, differential water level alarm should be determined. The high-level alarm will alert the operator that the screen is dirty and should be cleaned. This set point should be based on the normal mean ERCW flow rate. The set point should be about three inches greater than the clean-screen differential level, or that corresponding to 50 percent clogging, whichever is greater.

The high-high differential level alarm is based on structural limitations and will alert the operator of impending screen failure. The structural design limit should be obtained from the responsible engineer. The high-high alarm set point should be half the structural limit.

5.0 SPECIAL PROBLEMS

5.1 AIR LIBERATION

The ERCW system employs an evaporative cooling device to expel the heat absorbed by the system to the atmosphere. Such devices necessarily bring the water into intimate contact with atmospheric air. Since the atmospheric gases (primarily nitrogen and oxygen) are soluble in water, significant quantities if these gases can be absorbed and carried into the piping system. In addition to the corrosion problems caused by the oxygen, the gases can also cause flow disturbances and reduced heat transfer rates if they come out of solution inside the system.

Henry's law relates the important variables affecting gas solubility as follows:

$$N_X = \frac{P_X}{K_X}$$

Where N_X = Mole fraction of dissolved gas x

P_X = Partial pressure of dissolved gas x

K_X = Henry's law constant for gas x

This equation shows that the number of moles of gas that can be dissolved in a given quantity of water is directly proportional to its partial pressure and inversely proportional to its Henry's law coefficient. For a particular gas, the latter coefficient is primarily a function of temperature, increasing as the temperature increases. Thus, water becomes less soluble to air as the pressure decreases or the temperature increases. Since this is exactly what happens as raw water passes through the ERCW piping system, it is possible for gases to be released within the system.

As already noted, air liberation within the system can affect heat transfer of system coolers. For example, if a cooler was operating with a sufficiently high temperature rise, or was located so that the local pressure was sufficiently low, air liberation could occur. This would not necessarily create any problems. However, if the geometry of the cooler or ERCW piping were such that liberated air could be trapped within the cooler, the air could displace the water and effectively reduce the available heat transfer area. This could occur in coolers with a bottom outlet or where the ERCW piping turned down at the outlet nozzle.

Air liberation can also affect fluid flow if conditions are such that the flow is in the so-called slug flow regime. In this regime, large bubbles of liberated gases are periodically swept through the system, so the ERCW flow is not steady but pulsing. Unacceptable vibration can also result.

The preferred approach in initial design is to avoid air liberation entirely. The first step to achieve this goal is to calculate the maximum quantity of air dissolved in the ERCW inlet water. Henry's law may be used in the following combined form, which includes the effect from all atmospheric gases:

$$n_{air} = \frac{P_b/760}{K_{air}} \times C$$

Where N_{air} = Mole fraction of all dissolved gases, $\frac{\text{Moles-air}}{\text{Total moles}}$

P_b = Barometric pressure, mm Hg

K_{air} = Henry's law constant from Figure 1

C = Saturation factor (see below)

The barometric pressure should be the standard atmospheric pressure corresponding to the elevation of the free surface water level.

To maximize N_{air}, the Henry's law constant, K_{air}, should be based on minimum ERCW supply temperature. This is usually assumed to be 35°F.

The final factor, C, accounts for the degree of saturation of the inlet water. Since it is common for rivers to approach 100 percent saturation in the winter, C should not be less than 1.0. However, it is also known that river water can be supersaturated with air downstream of dams. Turbine operation can cause saturation up to 100 percent, while spillways can cause an increase to 150 percent. Evaporative devices such as cooling towers or spray systems may therefore justify a C value of 1.5, whereas a cooling lake may dictate a value of 1.0.

Having determined the mole fraction of dissolved air in the ERCW inlet water, the next step is to determine the minimum pressure required to keep this air in solution. Again, use Henry's law:

$$P_{reqd} = 760 \frac{N_{air}}{K'_{air}}$$

Where P_{reqd} = Minimum pressure to keep air in solution, mm Hg

      N_{air} = Maximum mole fraction in the ERCW inlet water

      K'_{air} = Henry's law constant based on cooler outlet temperature

The key difference here is that K'_{air} is based on cooler outlet temperature, T_{out}. The outlet temperature may be calculated using the following equation:

$$T_{out} = T_{in} + \frac{Duty}{500 \times Q}$$

Where T_{out} = Cooler outlet temperature, °F

      T_{in} = Cooler inlet temperature, °F

      Duty = Maximum cooler heat load, BTU/hr

      Q = ERCW cooler flow rate, gpm

Since T_{out} is greater than T_{in}, K_{air} will be lower than K'_{air} as evident from Figure 1. T_{out} must be calculated for each cooler and P_{reqd} be determined for each different value.

The next step is to determine the actual minimum pressure, P_{actual}, on the discharge side of each cooler. This simply involves calculating the head loss using Mechanical Design Standard DS-M3.5.1 and applying the energy equation. Static head is particularly important here since this usually predominates on the discharge side of the system beyond the flow-balancing valve.

If P_{actual} is less than P_{reqd} the potential exists for air liberation. As stated earlier, the fact that air is released is not necessarily unacceptable. The flow regime must be determined to make this evaluation. This can be done with the aid of Figure 2.

To evaluate those coolers having the potential for air liberation, it is necessary to first calculate the Froude number, F_r:

$$F_r = \frac{V^2}{gD}$$

Where V = Local velocity, ft/sec

g = 32.2 ft/sec²

D = Pipe inside diameter, ft

Next calculate the air-to-water volume ratio, V_r, of air liberated at the pipe section being investigated. To determine this ratio, it is first necessary to find the actual mole fraction of air released. This will be the difference between the mole fraction of dissolved air in the inlet water, N_{air}, and the mole fraction that local conditions will permit to remain in solution, N'_{air}.

$$N'_{air} = \frac{P'}{14.7 \, K'_{air}}$$

Where   P' = Local static pressure, psia

K'$_{air}$ = Henry's law constant at the local temperature (from Figure 1)

The volume of air liberated per unit volume of water, V_r, is then calculated by the following equation:

$$V_r = \frac{(N_{air} - N'_{air})(T')(1545)}{144 \, P'}$$

Where V_r = Air-to-water volume ratio

T' = Local temperature,

P' = Local static pressure, psia

298

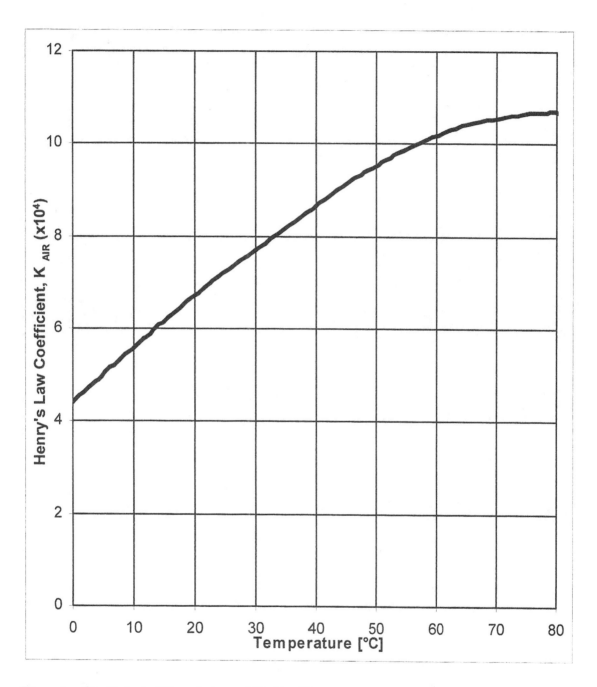

Figure 1   Coefficient of Henry's Law at Various Temperatures

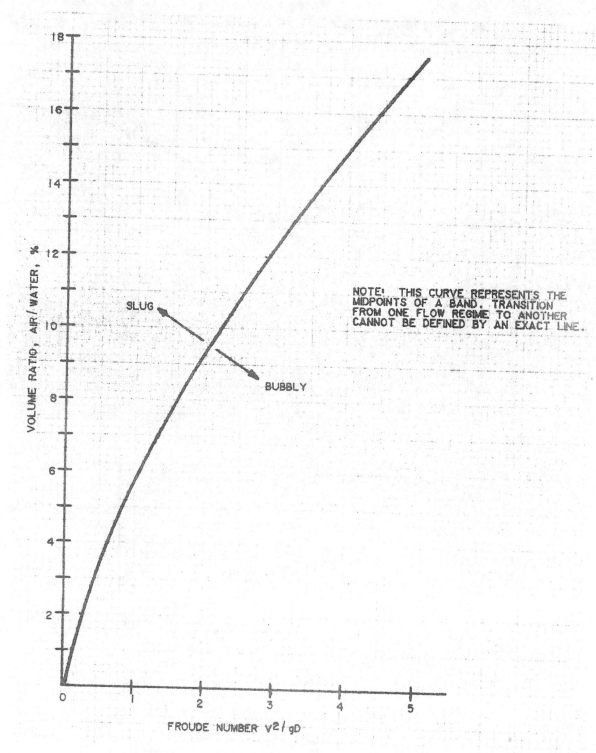

NOTE: THIS CURVE REPRESENTS THE MIDPOINTS OF A BAND. TRANSITION FROM ONE FLOW REGIME TO ANOTHER CANNOT BE DEFINED BY AN EXACT LINE.

Figure 2. Bubbly/Slug Flow Regime Map

The values of F_r and V_r are then plotted on Figure 2. Those
that fall within the bubbly flow regime are generally acceptable.
Those in the slug flow region may require modification.

Many possibilities exist to correct cooler loops found to be in the
slug flow regime; however, many of the solutions become increasingly
difficult as design and construction progress. Perhaps the simplest
solution is to relocate the flow-balancing valve beyond the region
having the potential for air liberation. This will pressurize the
critical area by the amount of the differential across the valve,
possibly eliminating the potential for air release or transforming
it to bubbly flow. Another possibility is to reroute the piping to
eliminate problem high points. It may even be possible to relocate
a cooler to a lower floor. Booster pumps to pressurize problem
areas are also effective, particularly on the supply side of the
coolers.

5.2 FLASHING/CAVITATION

5.2.1 Flashing

The term "flashing" refers to the change of state from liquid to
gas. It is generally associated with the formation of large
volumes of vapor as a result of heating, expansion into a low
pressure region, or both. Flashing might occur, for example,
within a cooler, because of the temperature rise of the water.
The effective heat transfer area would thereby be reduced, leading
to a further rise in temperature, more flashing, etc. Eventually
the cooler could become completely filled with vapor and unable to
remove its design heat load.

Flashing can also occur in regions of low pressure. Such regions
might exist downstream of flow-balancing valves or automatic
temperature control valves. For this reason, such valves should
generally be located downstream of their associated coolers to
preclude the possibility of collecting vapor within the cooler.

The first step in evaluating the potential for flashing is to
determine the cooler outlet temperature:

$$T_{out} = T_{in} + \frac{Duty}{500 \times Q}$$

Where T_{out} = Maximum cooler outlet temperature, °F

T_{in} = Maximum cooler inlet temperature, °F

Duty = Maximum cooler heat load, BTU/hr

Q = ERCW flow rate, gpm

The saturation vapor pressure, P_{sat}, corresponding to T_{out} may then be read from the steam tables. Next, the local cooler outlet pressure is calculated as follows:

$$P_{local} = [H_{pump} - h_L - (E_{local} - E_{supply}) - \frac{V^2_{local}}{64.4}] \frac{1}{2.31} + P_{atm}$$

Where P_{local} = Local static pressure, psia

H_{pump} = Total head of ERCW pump, feet of water

h_L = Head loss from pump to cooler outlet, feet of water

NOTE

Head losses should be calculated using procedures from Mechanical Design Standard DS-M3.5.1. Friction losses should be based on the minimum design C-factor.

E_{local} = High-point elevation of cooler outlet, feet MSL

E_{supply} = ERCW supply-side free surface water elevation, feet MSL

V_{local} = Local ERCW velocity, FPS

P_{atm} = Barometric pressure corresponding to E_{local}, psia

If $P_{local} \leq P_{sat}$ flashing will occur.

The outlet of flow-balancing valves or temperature control valves should also be checked for flashing. The local temperature and corresponding vapor pressure are determined using the foregoing formulas. The local pressure, however, should be determined from the following equation:

$$P_{local} = [H_{int} + h_L - (E_{local} - E_{int}) - \frac{V^2_{local}}{64.4}] \frac{1}{2.31}$$

Where h_{int} = Cooling system interface pressure, feet of water (refer to subsection 3.2.2)

h_L = Head loss from valve outlet to cooling system interface, feet of water.

302

NOTE

> Head losses should again be determined using
> DS-M3.5.1, but with frictional C-factors for
> new pipe.

E_{local} = Valve outlet elevation, feet MSL

E_{int} = Cooling system interface elevation, feet MSL

Again, if $P_{local} \leq P_{sat}$, flashing will occur.

Flashing should not be allowed to occur within system coolers.
Possible corrective measures include all of those mentioned in
subsection 5.1 for air liberation. It might also be possible to
prevent flashing by increasing the design flow rate of the ERCW,
thereby decreasing the temperature rise. Throttling the ERCW flow
rate downstream of the cooler may also prevent flashing by increasing
the local cooler pressure. Cooler performance should be evaluated
when determining the feasibility of the latter options.

Flashing downstream of throttling valves may lead to unacceptable
cavitation, vibration, or surging. Criteria for evaluating this
problem are presented in the next section. If found to be unacceptable, and if the local temperature rather than local pressure is the
main problem, bypassing a portion of the ERCW cooler inlet flow to
mix with valve inlet flow might effectively prevent flashing.

5.2.2 Cavitation

Whereas flashing deals with the formation of vapor, cavitation deals
with the damaging effects as the vapor bubbles change back to
liquid. The implosion of the vapor cavities can generate extremely
high local pressures. The shock wave itself is sufficient to
fracture the grain boundaries of metal surfaces, causing removal of
whole grains and leaving a roughened metal surface. Erosion may
lead to structural failure or leakage. Corrosion accelerates the
process.

Cavitation is generally associated with throttling valves, orifice
plates, venturis, and other such devices that create a sudden change
in local pressure. The propensity for and severity of cavitation
may be determined by calculating the "cavitation index":

$$K_i = \frac{P_d - P_{sat}}{P_u - P_d}$$

Where K_i = Cavitation index, dimensionless

P_u = Local inlet pressure, psia

P_d = Local outlet pressure, psia

P_{sat} = Saturation vapor pressure at maximum operating temperature, psia

The K_i value so calculated may then be interpreted from Table 1, Figure 3, or Figure 4. Other data may also be used if more applicable. If unacceptable cavitation is indicated by this procedure, the corrective actions in subsections 5.1 for air liberation and 5.2.1 for flashing may be effective. Generally, the low outlet pressure is the major contributor to the cavitation problem, and steps to increase it will be effective. Adding a multihole restricting orifice downstream of the cavitating component will increase the backpressure of the component and reduce or eliminate the cavitation. However, the effect on system performance at off-design points and over the plantlife must be carefully considered. If other measures are not feasible, the use of a special anti-cavitation valve or multiple orifice might be considered. Initial piping design should attempt to minimize the required throttling in the high-flow cooler loops.

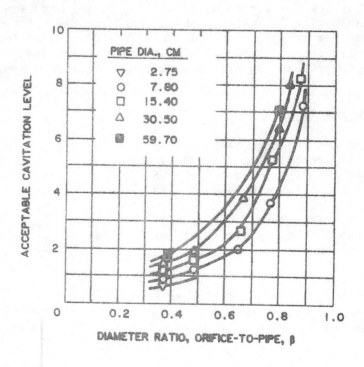

Source: Tung, Patrick C, and Mileta Mikasinovic, "Eliminating Cavitation From Pressure-Reducing Orifices," Chemical Engineering, Dec. 12, 1983.

Figure 3. Cavitation Indices for Orifices

TABLE 1. CRITICAL CAVITATION INDICES

Valve Type	Valve Opening as Percentage	Critical Cavitation Index[a]
8-inch cone[b,e]	30	1.40
	40	2.20
	50	2.75
	60	3.20
	70	3.50
	80	4.25
8-inch cone[b,f]	50	3.25
	60	3.35
	70	4.15
14-inch cone[b,e]	20	1.25
	30	1.55
	40	2.25
	50	2.80
	60	3.25
	70	3.55
24-inch cone[b,e]	5	0.60
	10	1.25
	20	1.90
	30	2.25
	40	2.50
8-inch ball[b,e]	10	1.75
	20	1.85
	30	2.00
	40	2.20
	50	2.50
	60	3.10
	75	4.50
8-inch ball[b,h]	18	1.85
	33	2.10
	44	2.30
	54	2.85
8-inch disc[b,g]	20	2.45
	30	2.80
	50	4.50
	58	6.00
8-inch disc[b,h]	20	2.65
	30	3.15
	45	4.40

305

TABLE 1. CRITICAL CAVITATION INDICES (Continued)

Valve Type	Valve Opening as Percentage	Critical Cavitation Index[a]
12-inch globe[b,e]	10	1.50
	20	1.75
	30	2.50
	40	3.40
	60	4.00
8-inch needle with 14-inch expansion[b,e]	8	0.40
	13	0.55
	23	0.70
	83	1.20
8-inch needle with 14-inch expansion[b,h]	8	0.40
	13	0.55
	23	0.80
	53	1.18
	83	1.50
8-inch needle with 12-inch expansion[b]	83	1.5[e]
	8	0.56[h]
	13	0.68[h]
	23	0.92[h]
	53	1.30[h]
	83	1.60[h]
8-inch needle with 8-inch expansion[b,e]	10	0.60
	30	1.20
	50	1.70
	80	2.10
12-inch ball with sudden expansion downstream[c,e]	20	2.50
	35	2.70
	50	3.15
	70	4.75
12-inch ball with sudden expansion downstream (air injected into side of pipe)[c,e]	20	0.60
	50	3.15
12-inch butterfly with sudden expansion downstream[c,g]	20	2.25
	35	2.00
	50	2.25
	70	2.90

306

TABLE 1. CRITICAL CAVITATION INDICES (Continued)

Valve Type	Valve Opening as Percentage	Critical Cavitation Index[a]
12-inch butterfly with sudden expansion downstream, (air injected into stem)[c,g]	20 50	0.60 0.75
6-inch butterfly[d]	17 32 50 66 88 100	0.90 1.10 1.36 1.90 2.20 1.60
6-inch gate[d]	12 25 38 50 62 68 75	1.10 1.10 1.40 1.80 1.60 1.30 1.00
6-inch globe[d]	20 40 60 80 100	0.80 0.60 0.51 0.45 0.45

a. The critical or incipient cavitation index is that at which cavitation effects are first noticeable.
b. Albertson, M. L., Tullis, J. Paul, and Thomas, Charles W., "Hydraulic and Cavitation Characteristics of Valves," Report No. 1, Hydro-Machinery Laboratory, Department of Civil Engineering, Colorado State University, Fort Collins, Colo., Vol. 2, June 1967.
c. Skinner, M. M., and Tullis, J. Paul, "Results of the Testing of a 12-Inch Ball Valve and a 12-Inch Butterfly Valve with a Downstream Expansion," Report No. CER60-67MMS-JPT26, Department of Civil Engineering, Colorado State University, Fort Collins, Colo., Jan., 1967.
d. Miller, E., and Huddleston, D., "Cavitation in Valves," Glenfield Gazette, Kilmarnock, Scotland, Vol. 35, No. 1, Spring, 1967.
e. 60-65 psi inlet pressure
f. 100-150 psi inlet pressure
g. 75-80 psi inlet pressure
h. 150 psi inlet pressure

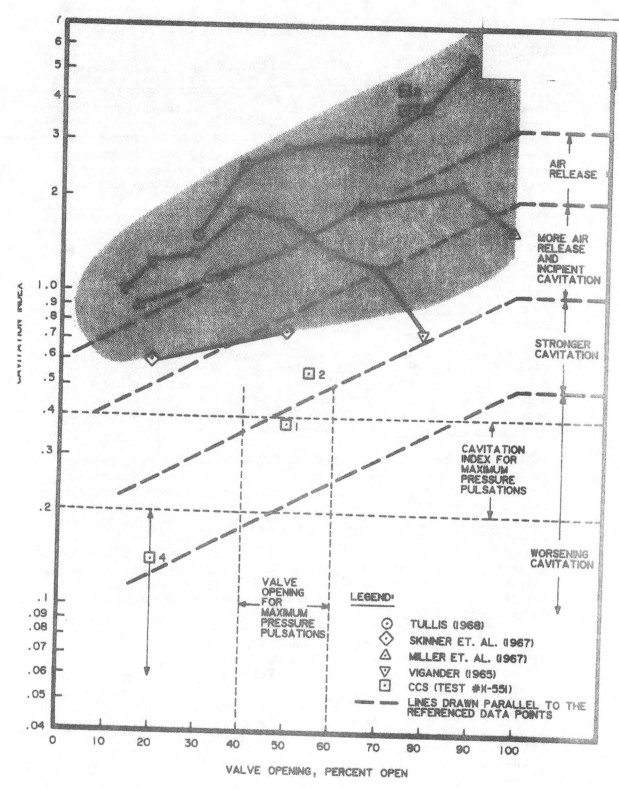

Source: Vigander, Svein, "Watts Bar Nuclear Plant -- Investigations of Cavitation in ERCW Throttling Valves," TVA Report No. WR28-1-85-107, Dec. 1983.

Figure 4. Cavitation Regions for Partially Open Butterfly Valves

5.3 AIR/VACUUM RELEASE

If the ERCW design uses vertical pumps, provisions shall be made for air and vacuum, release between the pump discharge and the discharge check valve. The vacuum relief valve is required when the difference in pump discharge elevation and the design low water level is greater than 30 feet to prevent column separation when a pump is shut down. When the pump is restarted, the air in the piping must be vented to prevent air from being forced into the ERCW system, which could degrade the performance of heat exchangers served by the ERCW system.

The air/vacuum release must be designed to allow air to be vented at the proper rate. If it is undersized, the air inside the pump column and pump discharge piping will not all be vented and some of it will be forced through the check valve into the ERCW system. If it is oversized, the air will be vented too rapidly, allowing the pump to run out and causing water hammer when the high velocity water column from the pump discharge hits the closed check valve with static water behind it.

The air/vacuum release system should consist of an isolation valve, an inverted check valve, a flow restricting orifice, and the required piping. The piping, isolation valve, and check valve should be approximately twice the diameter of the orifice. This will allow the orifice to be the source of most of the pressure drop and thus more accurately control the air flow rate.

The air release function of the air/vacuum release system requires greater capacity in terms of volumetric flow rate than the vacuum release function. For this reason, the orifice should be sized based on the air release requirements. The first step in determining the required size of the orifice is to determine the air mass flow rate required through the orifice. This requires knowing the system pressure against which the pump will be started and the pump flow rate at that pressure. The air density at this pressure must also be determined. Assuming the pipe is initially filled with ambient air before the pump is started, the density at system pressure is determined as follows:

$$P_c \ (\text{lbm/ft}^3) = P_a \ (\text{lbm/ft}^3) \left(\frac{P_c \ (\text{psia})}{P_a \ (\text{psia})} \right)^{1/1.4}$$

Where P_a = initial density (at 14.696 psia and 59°F P_a = 0.07651 lbm ft^3)

P_a = initial pressure

P_c = system pressure

From this information the air mass flow rate required can be determined as follows:

$$m \ (\text{lbm/sec}) = \frac{Q \ (\text{gal/min}) \ x \ P_c \ (\text{lbm/ft}^3)}{7.4805 \ (\text{gal/ft}^3 \ x \ 60 \ (\text{sec/min})}$$

Where Q = the pump flow rate at system pressure

P_c = air density at system pressure

m   = air mass flow rate in lbm/sec

Assuming that the system pressure is at least twice the ambient pressure, the flow through the restricting orifice will be very near sonic velocity.

The ratio of air density of sonic conditions to the stagnation density, $P^*/P_{\text{stagnation}}$, is 0.6339, and the ratio of the speed of sound at sonic conditions to the speed of sound at the stagnation conditions, $a^*/a_{\text{stagnation}}$, is 0.9129. In this case the stagnation conditions are the pressure, temperature, and density of the air in the pipe after it has been compressed by the pump start. The speed of sound is a function of temperature and the temperature of the compressed air can be determined as follows:

$$T_c = T_a \ (P_c/P_a)^{.4/1.4}$$

Where T_a = Absolute temperature of the air initially in the pipe

P_a = Pressure of air in pipe before pump start

P_c = Pressure of compressure air (system pressure)

T_c = Absolute temperature of compressed air

The orifice plate should be located downstream of the isolation valve and check valve. This will ensure that choked flow will occur at the orifice. Assuming that approximately 10 percent of the available pressure drop would occur in the two valves upstream of the orifice, the stagnation pressure at the orifice would be reduced by 10 percent of the available pressure differential. The stangation temperature would remain unchanged. The speed of sound at this condition can be determined by solving the following equation.

$$a_{\text{stagnation}} \ (\text{ft/sec}) = 49.1 \ \sqrt{T_{\text{stagnation}} \ ^\circ R}$$

310

The density at stagnation conditions at the orifice entrance with the 10 percent reduction in pressure differential would be

$$P_{stagnation} = P_c \left(\frac{\left((P_c - P_a)x \; .9\right) + P_a}{P_c} \right)^{1/1.4}$$

The size of an opening that will pass the required mass flow rate can now be determined by solving the following equations:

$P^* = P_{stagnation} \; x \; P^*/P_{stagnation}$

$V^* = a^* = a_{stagnation} \; x \; a^*/a_{stagnation}$

Where $V^* =$ velocity through the opening (ft/sec)

The mass flow equation is written

$m = P^* \; V^* \; A^* = P^* \; V^* \; d^{*2}/4$

Solving for d^* in inches

$d^* (in.) - 24 \; x \; \sqrt[4]{m/(\Pi P^* \; V^*)}$

From this the orifice should have a diameter of d^* and the piping, isolation valve, and check valve should have a diameter of approximately $2d^*$. The 10 percent pressure drop assumed is an approximation based on having pipe twice the diameter of the orifice and valve throat diameters of 90 percent of the pipe inside diameter. If in operation, the pressure drop is less than assumed, this results in water hammer and the isolation valve can be throttled slightly.

During preoperational testing, the system shall be closely observed and instrumented to detect any vibration or pressure spikes which might indicate the occurrence of water hammer, especially during pump startup and shutdown. Any required changes in the orifice size or throttling of the isolation valve can then be made.

Chapter 9
Design Guide DG-M3.1.4: Closed Cycle Ultimate Heat Sink

1.0 GENERAL

1.1 EFFECTIVITY

This design guide presents procedures and techniques for analyzing the essential raw cooling water (ERCW) system ultimate heat sink (UHS) in accordance with the requirements of the Nuclear Regulatory Commission (NRC) Regulatory Guide (RG) 1.27. This design guide applies only to ERCW systems that use a closed cycle evaporative cooling device such as a spray pond, mechanical draft cooling tower, or cooling lake.

Before beginning the UHS analysis, appendix 9.2A of the Safety Analysis Reports for previous nuclear plants should be reviewed as examples.

1.2 DEFINITIONS/ABBREVIATIONS

CEB	Civil Engineering Branch
CWT	Cooled water temperature
ERCW	Essential raw cooling water
ES	Emergency shutdown
FSAR	Final Safety Analysis Report
LOCA	Loss of coolant accident
LOP	Loss of offsite power
MEB	Mechanical Engineering Branch
met.	Meteorological
NCC	National Climatic Center
NEB	Nuclear Engineering Branch
NEG	Nuclear Engineering Group
NGG	Nuclear Steam Generation Group
npsh	Net positive suction head
NRC	Nuclear Regulatory Commission
NSA	Nuclear Safety System Group
NSR	Nuclear Steam Generation and Radiological Group
NSSS	Nuclear Steam Supply System
PSAR	Preliminary Safety Analysis Report
RG	Regulatory guide
RI	Reactor island
RSI	Ryznar stability index
UHS	Ultimate heat sink

2.0 SYSTEM REQUIREMENTS

Before analyzing the ERCW system, the engineer should become thoroughly familiar with NRC requirements, system design features, and the requirements of the components served by the system. These must be evaluated concurrently in order to determine the worst-case conditions for analysis.

313

2.1 NUCLEAR REGULATORY COMMISSION

The NRC requirements are stated in the latest revision of NRC RG 1.27. The term "ultimate heat sink" (UHS), as used therein, means the entire ERCW system, including all backup equipment for the various modes of operation. The engineer should study NRC RG 1.27, "Ultimate Heat Sink for Nuclear Power Plants," in order to apply it properly to the ERCW system. For convenience, the requirements of NRC RG 1.27 Revision 2, which affect the UHS analysis, are summarized in the following paragraphs.

The UHS must have a sufficient inventory of water to ensure safe shutdown of the plant for 30 days after shutdown initiation without any makeup from outside the UHS.

The UHS must be capable of removing heat from the reactor and its auxiliaries at a rate that is sufficient to permit safe shutdown of the plant. [Compliance with this requirement is often shown by demonstrating that the temperature of water returned from the UHS to the reactor island (RI) does not exceed a design basis maximum value.]

The UHS must have sufficient inventory and heat transfer capability to perform its function adequately in spite of a single failure; severe natural phenomena such as earthquake; extreme meteorological (met.) conditions; unusual plant conditions such as loss of coolant accident (LOCA) or emergency shutdown (ES).

NRC RG 1.27 should be consulted for a complete and precise statement of these requirements.

2.2 SYSTEM DESIGN

The engineer should consider the availability of all portions of the system for service under the various postulated upset conditions and failure modes. The analyses should be compatible with the system design bases as described to the NRC in section 9.2 of the Preliminary Safety Analysis Report (PSAR) or Final Safety Analysis Report (FSAR). The capability of different parts of the UHS to function under various sets of conditions is a key factor in determining the conditions to be analyzed. For example, certain parts of the UHS may be unavailable following an earthquake or tornado, and other parts must then be available to fulfill the role. Flexibility in the operating modes of the system components is another key factor. For example, shutting off or bypassing certain components may reduce the system heat load or the rate of water inventory loss.

2.3 COMPONENTS SERVED BY ERCW

The flow rate and heat load requirements for individual components served by the ERCW system under each of the postulated sets of conditions should be considered. Examples of things to be considered are whether nonessential components are to be served after loss of offsite power (LOP), whether one division (or train) may be turned off in order to reduce the requirements on the UHS, the variation in component heat load as a function of time, maximum allowable ERCW temperature supplied to the components, limitations on water quality, and limitations on velocity of water through the components. If the reactor auxiliaries are designed under outside contract, most of this information should be available in documents supplied by the contractor.

The engineer should work closely with the Nuclear Safety System Group (NSA) and the Nuclear Steam Generation and Radiological Group (NSR) of the Nuclear Engineering Branch (NEB) to ensure that the component requirements are properly identified and applied.

3.0 THERMAL ANALYSIS

3.1 DESIGN BASIS CONDITIONS

3.1.1 Flow Rate and Heat Load

The required flow rate and heat load as functions of time may be determined for each possible mode of operation from the system design and the component requirements discussed in sections 2.2 and 2.3. The flow and heat load determine the cooling range needed. It will usually be possible to determine by examination of the flow, heat load, range, and thermal inertia (or response time) of the UHS that only one or two operating conditions could possibly represent the worst cases for heat transfer. If it is not clear which condition is most severe, then an analysis of each likely condition can be made in order to identify the worst case. It will normally be sufficient for these analyses to use approximations of the final input data, because slight changes in input data would not change the relative severity of the conditions.

3.1.2 Meteorological Data

Revision 2 of NRC RG 1.27 does not specify a detailed basis for selection of met. data for use in analysis. It does identify factors which must be addressed in the selection process. The requirements have sufficient flexibility to permit the selection process to be tailored to the characteristics of the UHS.

3.1.2.1   Sources of Data

Met. data recorded at numerous locations over long periods of
time is available on magnetic tapes from the National Climatic
Center (NCC). It is generally possible to find such data for a
location in the region of the plant and suitable for use in the
analysis. Data on solar radiation is not recorded on these
tapes, but is available from the NCC in other forms. Values for
solar radiation may also be obtained from various reference
books or calculated as a function of latitude and season.

3.1.2.2   Selection of Worst Case Data

UHS performance under various met conditions must be approximated
mathematically in order to identify worst-case met. data. A
simple expression for performance as a function of met. data
(such as a curve fit or an explicit equation) is necessary
because of the vast number of met. data points for which
calculations will be carried out. For example, the cooled
water temperature (CWT) of a spray pond under steady state
conditions may be evaluated as a function of nozzle efficiency,
required cooling range, wet bulb temperature, and wind speed.
The CWT of a cooling tower might be evaluated using a curve
of CWT as a function of wet bulb temperature for the required
cooling range. The worst-case met. data is considered to be
that which results in the highest average CWT over 30 consecutive days based on the chosen relationship for CWT.

As the design of the ERCW system develops, there may be changes
in the functional relationship between CWT and met. conditions.
If these changes are major, then the revised relationship should
be used to make a new determination of worst-case met. data. It
has been found, however, that minor changes in the relationship
often result in no change to the selected met. data.

Since NRC RG 1.27 calls for analysis of a 30-day period, it is
necessary to establish a 30-day criterion, using system conditions typical of the first 30 days of shutdown. The met. data
records should be searched[1] to find the 30-consecutive-day
period which has the highest average CWT.

---

1. The details of the searching procedure are beyond the scope of this
    document. Additional information on a suitable procedure may be found
    in Mechanical Design Guide DG-M3.5.4, "Evaluation of Recorded Meteorological
    Data."

Revision 2 of NRC RG 1.27 also calls for the use of met. data appropriate to the critical time period(s) for the UHS. The phrase "critical time period" is interpreted as referring to the amount of time required for the temperature of the UHS to reach a peak in response to some change in operating conditions (i.e., response time), although NRC RG 1.27 uses the phrase ambiguously. The length of this critical time period may be determined from appropriate analytical solutions, if available. It may also be determined by performing a 30-day transient analysis with identical met. conditions for all days and observing the response of CWT. (Note that diurnal variations in met. conditions should be used in such an analysis.) The critical time period is typically on the order of 1 to 5 times the recirculation time of the ERCW system. The worst-case met. data for the critical time period should be determined by searching the met. data records using an appropriate criterion based on system conditions typical of the critical time period.

For some types of UHSs, more than one critical time period would be appropriate. In such cases, appropriate criteria should be developed and applied to determine worst-case met. data for all critical time periods.

For example, the response time of a spray pond decreases as pond inventory is depleted during the shutdown period, which results in a tendency for pond temperature to peak not only in the first days of shutdown but also at the end of the 30-day period. Another example is given in the discussion of a cooling lake in section B of NRC RG 1.27, but it is ambiguous.

3.1.2.3   Arrangement of Worst Case Data

NRC RG 1.27 allows flexibility in the application of worst case met. data, yet makes it clear that such application is to be done conservatively. It does not specify whether the data for the worst-30-day period should be used in chronological order as recorded, or whether the data should be averaged to construct a synthetic day representative of the entire period, or some other method is to be used. To insure conservatism, NRC RG 1.27 requires that during any critical time period, met. data appropriate for that period be used. Whenever CWT is at its peak, conservative met. data (e.g., worst-1-day) must be used.

It is convenient to represent the met. data for any period of more than one day by data for a synthetic day that is obtained by averaging the data for the entire period. This eliminates random fluctuations in CWT that would result from day to day variations in met. data, giving a clearer picture of trends in

317

the system performance. It also simplifies the preparation of
input data for computer programs. Since worst case data for
periods of one day is to be inserted immediately before, during,
and after the CWT reaches a maximum, the maximum calculated CWT
will be conservative.

3.2 ULTIMATE HEAT SINK PERFORMANCE

A mathematical model of UHS thermal performance, which predicts CWT
as a function of system operating conditions and met. conditions, is
the obvious key to the entire thermal analysis. Such a model should
give conservative estimates of CWT so that there is confidence in
the results. If the conservatism is excessive, the model would
require an unnecessarily expensive or even impractical UHS design.
If the UHS is designed by a contractor, then the model should be
based upon the contractor's guaranteed performance data. A suitable
procedure for calculating the thermal performance of spray systems
is described in Mechanical Design Guide DG-M3.1.6, "Spray System
Thermal Performance Calculations." Further discussion of the UHS
performance model depends on the type of UHS employed and is beyond
the scope of this design guide.

3.3 ANALYSIS

The final thermal analysis will generally be performed using a
computer program which predicts the CWT of the UHS in response to
the design basis flow rate, heat load, and met. data. Certain steps
must be taken prior to the final analysis. These steps are likely
to involve preliminary computer analyses and serve to determine
input data for the final analysis.

3.3.1 Initial Conditions

NRC RG 1.27 does not specifically identify requirements for the
determination of the UHS temperature at the time of initiation of
shutdown. It should be obvious, however, that a conservative
determination must be made. This determination should be based
upon worst-case operating conditions prior to shutdown and upon
severe met. conditions. The met. conditions that are used prior
to shutdown and the length of the preshutdown period that is
analyzed should be chosen to provide a realistic worst-case
temperature at the time of initiation of shutdown. It would be
appropriate to identify a preshutdown critical time period and to
use worst-case met. data selected for this period.

If the design of the UHS is such that the CWT peaks soon after
shutdown begins, then the maximum value of CWT will depend
upon the time of day that shutdown begins. In such cases it

318

is necessary to synchronize the meteorology with the shutdown heat loads in order to obtain the maximum CWT. In some cases this may be done by inspection. In others, it is necessary to perform preliminary analyses with shutdown beginning at varying times during the day and to select the worst case based on the results.

3.3.2 Changing Conditions

Any changes in conditions that would occur during the period of analysis must be modeled in such a way that the effects upon UHS performance are conservatively predicted. If action by the unit operator is required to assure UHS performance, then sufficient time must be provided for the operator to recognize the need for the action and to see that it is accomplished.

Examples of such changes may be seen by considering a UHS that has an active water inventory that circulates through the RI, and an inactive water inventory that replenishes losses from the active inventory. The performance of the sink may be adversely affected by losses from the active inventory. The analysis of this UHS should reflect the declining active inventory. After sufficient time has elapsed for the operator to recognize and alleviate the need for transfer of water from the inactive inventory to the active inventory, the analysis may reflect that such transfer has begun. The analysis should use a conservatively determined value for the temperature of the transferred water.

3.3.3 Margin

It is usually impractical to perform an analysis which is clearly conservative in every respect, especially since the system design may not be finalized until after the analysis is completed. Consequently, the analysis should use reasonable conservatism wherever practical and should reflect an adequate margin to cover uncertainties. The amount of margin is the amount by which the ERCW design maximum supply temperature exceeds the maximum temperature calculated in the UHS analysis. The amount of margin required may be found in the ERCW system design criteria document and should be at least 2°F.

4.0 INVENTORY ANALYSIS

4.1 TYPES OF INVENTORY LOSS

The analysis must consider all credible types of water losses under the postulated modes of operation. The following list of types of water loss is representative but may not be complete.

4.1.1 Losses Prior to the 30-Day Period

Sedimentation may occur in the UHS, causing part of the volume of the UHS to be occupied by sediment with a corresponding loss of water inventory.

Consideration should be given to various operating procedures, unusual events, and failure modes (such as loss of makeup) in order to determine the credibility of a partially-depleted inventory when the 30-day period begins.

4.1.2 Losses During the 30-Day Period

Water is lost from the UHS throughout the 30-day period as a direct result of UHS operation to dissipate heat. Such losses include evaporation and drift. They typically represent over half of the total inventory losses.

The UHS will dissipate most of the heat loads by evaporation, with only a small part being dissipated by sensible heat transfer. It is convenient to identify the evaporative losses with the source of the heat load. The largest evaporative losses result from the RI heat loads. The design of the UHS may be such that there are no other significant heat loads. However, if the UHS has a large surface area (e.g., spray ponds or cooling lakes) then the solar radiation to the surface of the UHS will result in significant additional heat loads and evaporation.

Spray ponds and cooling towers lose some water as drift. Drift consists of those small drops which are carried out of the cooling device by the air currents which flow out of the device.

A substantial amount of inventory may be lost in ways that are unrelated to the UHS primary function of heat dissipation. Some of these losses are inherent in the design of the UHS, such as seepage from the reservoir, leakage from the piping, and blowdown. Other losses may result from requirements that the UHS provide inventory for secondary functions, such as essential fire protection and makeup to certain essential components or systems.

4.1.3 Inventory Required After 30 Days

Some inventory should remain at the end of the 30-day period to ensure that the UHS can still perform adequately at that time. All possible constraints should be considered in order to determine the minimum allowable inventory at this time. Examples of such constraints are net positive suction head (npsh) for the ERCW pumps, water quality (e.g., scaling or corrosion tendency and chloride content), and possible erosion of exposed spray pond liner material as a result of direct impact of sprayed water upon the liner. Note that these requirements are not additive. A remaining inventory, which is adequate for the most severe constraint, is also adequate for the less severe constraints.

4.1.4 Contingency

The actual inventory of the UHS should exceed the calculated inventory requirement by some margin in order to provide protection against changes in system requirements or other contingencies. There is no requirement in NRC RG 1.27 for such margin or contingency allowance. Instead, the contingency allowance is made up of that inventory that is left over after NRC RG 1.27 has been satisfied.

4.2 PROCEDURE FOR ANALYSIS

The inventory problem involves the integrated water requirements over the entire 30-day period, as contrasted with the thermal problem which involves instantaneous temperature and heat transfer rate. Hence, it is generally possible to address the inventory problem by considering only 30-day total needs, rather than instantaneous loss rates.

The various types of inventory requirements are generally independent of each other. Even the variables upon which the requirements depend are unrelated in most cases. (Some exceptions will be identified in the following sections.) It is thus feasible to analyze each type of inventory requirement separately and sum the calculated requirements to obtain the total required inventory.

Suitable procedures are described in the remainder of this section for determining the size of each inventory requirement which would normally have to be considered. No attempt will be made here to address unusual requirements which might occur because of special UHS design features.

4.2.1 Sedimentation

The estimated volume of sediment that might build up should be obtained from the Civil Engineering Branch (CEB). Its determination is outside the scope of this design guide.

321

4.2.2 Evaporation due to RI Heat Loads

This evaporation may be estimated from the integrated RI heat load
by assuming that all the heat is dissipated by evaporation. This
approach is generally accepted as conservative. The 30-day
integrated heat load for each postulated mode of operation should
be investigated to determine which mode results in the greatest
water loss from RI heat loads. It is likely that the mode thus
identified will result in the highest total inventory losses
of all types. However, the analyst should investigate the
possibility that some other mode will result in higher drift
loss and thus higher total inventory losses of all types.

4.2.3 Evaporation due to Solar Heat Loads

This loss, which may be referred to as solar evaporation, may be
calculated assuming that the entire solar heat load is dissipated
by evaporation. The intensity of solar radiation may be obtained
from the sources referred to in section 3.1.2.1.

Solar evaporation and drift are both dependent upon meteorology in
the case of spray ponds. The maximum solar evaporation is likely
to occur in midsummer, whereas the maximum drift is likely to
occur in springtime. If solar evaporation and drift are maximized
independently and then added together, the result is conservative
but may be excessively conservative. If this is found to be the
case, then some less conservative approach may be used. Care
should be taken to insure that the approach used does not render
the combined solar evaporation and drift loss estimates
nonconservative.

The coupling of solar evaporation and drift losses through
meteorology is not a problem with cooling towers and cooling lakes
because at least one of the two losses is negligible in each of
these cases.

4.2.4 Drift

4.2.4.1 Selection of Worst Case Meteorological Data

The procedure for selection of worst case met data for drift
analysis is quite similar to the procedure described in
section 3.1.2 for the thermal analysis. In this case, however,
the criterion to be used is an estimate of drift loss rate as
a function of met. data. The parameter of concern is the total
drift loss during the 30 day period, so the worst-case 30-
consecutive-day period is the only period for which worst-case
met. data is needed. As with the worst case met. data for thermal

322

performance, it is not considered necessary to reselect the worst-case data for drift as a result of minor changes in the design or characteristics of the UHS.

4.2.4.2 UHS Drift Characteristics

For mechanical draft cooling towers, the drift rate may be obtained from the manufacturer's guaranteed maximum drift rate, which is often expressed as a percentage of tower flow rate. It is usually considered to be independent of met. conditions.

For spray ponds, the drift loss rate is typically a function of met. conditions (primarily wind speed), spray system geometry, and the sprayed water flow rate.

If the spray system design is purchased, the drift rate characteristics may be obtained from the supplier. If suitable manufacturer's data is not available, use the procedures for calculation of spray system drift loss rate in EN DES Design Guide DG-M3.1.5, "Spray System Drift Calculations." These procedures should be applied conservatively to the specific design being considered, taking into account any significant differences in the location, orientation, design, and operating conditions of individual nozzles.

4.2.4.3 Total 30-Day Drift Loss

The UHS drift characteristics will normally be expressed as drift loss rate as a function of wind speed, with the function being a constant in certain cases. The total drift loss during the 30-day period is obtained by integrating the drift loss rate over the 30-day period, with the percentage drift rate at any time being determined by the wind speed at that time. The integral may be approximated by using a conservatively high constant wind speed, if a wind speed can be identified, which is clearly conservative yet does not result in excessive total drift loss. Another possibility is the use of a constant wind speed which is equal to the average wind speed for the 30-day period. The result is a rough approximation of the actual value, but is nonconservative because of nonlinearity in the drift vs. wind speed relationship. A much better approximation may be obtained by using daily average wind speeds. This may also yield slightly nonconservative approximations because of nonlinear drift characteristics, but in this case the effect should be negligible. A more accurate value of integrated drift loss could be obtained if hourly or trihourly values of wind speed were used. However, this refinement would be well within the accuracy of the estimates of drift rate and would not be justified.

4.2.5 Blowdown

If blowdown continues after the start of the 30-day period under the design basis conditions, it will substantially reduce the pond inventory. Because of this, the UHS may be designed to prevent blowdown if makeup is lost or if the water level in the reservoir drops below a certain level. The analysis must consider whether loss of water by blowdown is credible and, if so, how much water could be lost.

4.2.6 Seepage

Seepage of water from the reservoir is a function of several factors, e.g., the porosity and geometery of the reservoir. The amount of seepage loss associated with a particular UHS should be obtained from CEB. The method of determining this loss is outside the scope of this design guide.

4.2.7 Leakage

Leakage from those components of the UHS other than the inventory storage reservoir is expected to be small because of the quality assurance procedures associated with a safety-related system. However, some leakage may be expected to occur from time to time. A sizeable leakage rate which continued for 30 days could have a major impact on inventory. Predicting the total leakage rate for the entire system is virtually impossible because of the large number of components which could leak and the uncertain size of each leak. There are presently no NRC guidelines which specifically address leakage as it affects UHS inventory. The following guidelines are based on engineering judgement.

The analyst should verify that the plant design has adequate provisions (e.g., frequent inspection or flow instrumentation) for detecting during normal operation any leakage which is large enough to significantly affect inventory over a 30-day period. He should also verify that operating procedures will result in the prompt correction of any such leakage which is detected. (Such verification may be done by proper input to the system design criteria document and the operating instructions, for example.) Under these conditions, it is reasonable to assume that no significant leakage exists at the start of the 30-day shutdown period.

It should further be assumed that no significant leakage develops during the 30-day shutdown period. Predictions should be made of the transient inventory in various possible operating modes (e.g., normal shutdown or emergency shutdown with loss of makeup). These predictions should be made available to the plant operator so that he can monitor the inventory depletion rate and readily verify whether the inventory is being depleted at an acceptable rate or if significant leakage is occurring.

4.2.8 Supply for Secondary Functions

The UHS may be required to provide water inventory for certain essential functions which are basically unrelated to its primary function of heat dissipation. Such functions may include essential fire protection and providing makeup water to certain essential components or systems. These requirements may be identified through review of the TVA ERCW system design criteria and vendor documentation and through close coordination among groups within MEB and NEB.

4.2.9 Water Quality

As the water inventory diminishes during the 30-day period, the concentration factor for total dissolved solids in the system increases. (Guidance for calculating the concentration factor is given in Mechanical Design Guide DG-M3.1.3, "Makeup and Blowdown Rates for Evaporative Cooling Systems.") The concentration factor at any instant depends upon the initial concentration factor and inventory, remaining inventory, and what part of the dissolved solids has been carried out of the system with drift, seepage, leakage, blowdown, or supply of secondary functions. It also depends upon when these various losses occur. It depends upon whether (and how) the contingency allocation is assumed to be lost or whether the contingency is assumed to remain in the system.

High concentration factors can cause two types of problems. The first is a high concentration of problem-causing substances. For example, under certain conditions a high chloride content can damage stainless steel. The second type of problem is an increase in the tendency for scale formation associated with changing alkalinity, pH, total dissolved solids, and calcium content. In each case, it can be difficult or impossible to quantify the rate of development of harmful effects during the 30-day period. Consequently, judgment must be relied upon in establishing a suitable design basis.

Procedures are given in Mechanical Design Guide DG-M3.5.2, "Evaluation of Scaling and Corrosion Tendencies of Water," for calculating the Ryznar stability index (RSI) for varying temperature and concentration factors. The RSI is an indicator of scaling tendency but does not predict the rate or extent of scaling. Because the concentration factor increases rapidly when the remaining inventory approaches its minimum value, it may be impractical to prevent the RSI from entering the region of heavy scaling tendency for the entire 30-day period. As a general guide, however, the water quality allocation should be sufficient to prevent the RSI from entering the region of heavy scaling tendency until near the end of the 30-day period.

The calculations of scaling tendency should be based upon reasonably conservative water quality data at the initiation of shutdown. The system water quality responds very slowly to changes in makeup water quality because it takes many days (or even several weeks) to replace the volume of water in the system with makeup. Thus it would be quite conservative to use the worst recorded combination of water quality parameters, and nonconservative to use annual average values.

It may be possible in some cases to show that no allocation for water quality is necessary. If a sufficient quantity of water is to be available after 30 days for other purposes (such as pump npsh), then it may also serve to limit water quality adequately. The tendency of the system to deposit scale on heat transfer surfaces before or during the 30-day period may be controlled by treatment with chemicals (such as scale inhibitors).

5.0 REFERENCES

The following documents are referenced in the text and are essential to performing the UHS[2] analysis:

5.1 Mechanical Design Guide DG-M3.1.3, "DESIGN CONCEPTS OF LOW PRESSURE SYSTEMS--Makeup and Blowdown Rates for Evaporative Cooling Systems"; TVA, Knoxville: 1981.

5.2 Mechanical Design Guide DG-M3.5.2, "DESIGN CONCEPTS OF LOW PRESSURE SYSTEMS--Evaluation of Scaling and Corrosion Tendencies of Water"; TVA, Knoxville: 1981.

---

2. Mechanical Design Guide DG-M6.3.3, "WATER SYSTEMS--Essential Raw Cooling Water System," provides additional insight into the UHS analysis.

5.3   Mechanical Design Guide DG-M3.5.4, "DESIGN CONCEPTS OF LOW PRESSURE
      SYSTEMS--Evaluation of Recorded Meteorological Data"; TVA,
      Knoxville: planned 1982.

5.4   Mechanical Design Guide DG-M3.1.5, "DESIGN CONCEPTS OF LOW PRESSURE
      SYSTEMS--Spray System Drift Calculations"; TVA, Knoxville: 1981.

5.5   Mechanical Design Guide DG-M3.1.6, "DESIGN CONCEPTS OF LOW PRESSURE
      SYSTEMS--Spray System Thermal Performance Calculations"; TVA,
      Knoxville: planned 1982.

5.6   Regulatory Guide 1.27, "Ultimate Heat Sink for Nuclear Power
      Plants," Revision 2, U.S. Nuclear Regulatory Commission; NRC,
      Washington: 1976.